The Amazing
WHEAT BOOK

by LeArta Moulton

published by
LM Publications
509 E. 2100 N.
Provo, UT 84604
801.374.1858

LeArta Moulton, The Author, BYU graduate and mother of four has given radio shows, demonstrations and lectures for the past 25 years in seminars workshops and church functions on the uses of wheat. Those who have tasted her endless variety of wheat dishes know she is an authority on gluten cooking and seasonings.

The use of natural foods has been part of the author's family tradition. She has dedicated herself to teaching others how to use wheat to its fullest. *The Amazing Wheat Book* will give you exciting and new insight into using this wholesome grain.

Feeling that every mother has an obligation to do all she can for her family, LeArta is also in demand for her "down to earth" advice on the use of edible and medicinal herbs.

FORWARD

With love and appreciation, I dedicate this book to my father and mother, Arthur Wm. and Clara Andersen, who have taught me the value of wholesome, natural foods from the time I was small; to my husband and family, who have been patient and willing to assist me over the past years in adapting wheat, with all its varied uses, to our daily diet.

A special thanks to my sisters Esther Dickey and Ruth Laughlin for inspiring me to build upon much of their basic work with wheat, to another sister Beth Evans for her ideas and encouragement in creating the basic muffin recipe and to the rest of my eleven sisters who have supported and experimented along with me.

Layout and Cover – Cochran Studios
Illustrations – Sherrell Stewart

PREFACE

The need for wholesome foods is becoming more apparent all the time. Cancer, arthritis, heart by-passes and many other related health problems do not just come to everyone as they are getting older. It's our choice by the way we eat and take charge of our body that determines our fitness, health and ability to fight diseases.

Preparing foods from whole grains, beans and vegetables and to present it in a way to please the whole family is quite a challenge. There are so many quick mixes, frozen meals, and fast foods on the market today that it is tempting to overlook the comparison in food value.

I have dedicated the time I use in preparing foods to search out quick, easy methods and recipes to give the cook an edge on preparing health giving, tasty meals. Using the ideas in this book will make it easy to treat your family to the tastes they like and still have nutritional meals.

One of the secrets in making basic foods appetizing and flavorful is the seasoning and flavoring. There are many seasoning mixes on the market today, but you will pay a high price and have no choice in the ingredients of sometimes too much salt and harmful chemical preservatives. You will find many favorite tastes and blends in the Seasoning section and will be delighted to find how convenient and inexpensive it is to make your own.

Change your life for the better. Take control of your health and enjoy the tastes that are delicious as well as nutritious. Keep a positive outlook on life, exercise, develop a life-style of eating wholesome foods and you will feel good about yourself.

It is my hope that the recipes you find in this book will give you the satisfaction of eating better, feeling better and preparing it in the quickest way possible.

CONTENTS

History of Wheat

Wheat (T. vulgare, in the grass family GRAMINEAE), has played in important part in the development of civilization and has been grown before the beginnings of recorded history. It is the principal cereal foods of an overwhelming majority of the worlds inhabitants. It has a range of cultivation, probably greater than that of any other crop, except rice in the Orient, and is being seeded and harvested somewhere in the world each month of the year. Active wheat breeding in worldwide research programs continue to seek for high yields and enhanced nutrition. "Triticale" is an example and is a hybrid of rye and wheat which makes it high in protein; however, it does not have the gluten-forming properties. Lysine is a limiting amino acid and efforts are being directed to produce varieties with a higher lysine content.

Nutritional Value

Wheat contains the important B vitamins thiamin, riboflavin and niacin and is a rich source of Vitamin E, along with many other minerals and vitamins. Vitamin C is not present unless the grain is sprouted. The protein content varies from 7 to 19 percent.

Soft wheat, although low in protein, is still very beneficial. Use this wheat for sprouting, cracked wheat, popped wheat snacks, a postum-like drink and for any recipe calling for flour, i.e. pancakes, pastries, cookies, etc. Soft wheat is not ideal for isolating the gluten as the yield is not high, nor does is make a good light bread. Added commercial gluten flour helps.

Vitamin E aids in aging retardation, anti-clotting factor, blood cholesterol reduction, blood flow to the heart, capillary wall strengthening, fertility, male potency, lung protection (anti-pollution), muscle and nerve maintenance. The absence, or even shortage, of vitamin E could cause a tendency to arthritis, cancer, heart disease, and cerebral hemorrhage. This is the one vitamin necessary for normal reproduction and lactation. Lack of this vitamin can cause miscarriage. Vitamin E is necessary for growth and to build resistance to cancer and other degenerative diseases.

Vitamin B increases ability to learn and to remember, builds resistance to disease, promotes a healthy heart and good digestion, strengthens nerves, aids normal reproduction and increases the ability for mothers to nurse, improves circulation, which strengthens skin, hair, eyes, ears and liver.

Know Your Wheat

Select a hard wheat (16-19% protein), when possible. If in doubt, conduct a test by making gluten. A yield of 2 cups of raw gluten from 7 cups of flour indicates a high protein wheat. When comparing quality between different types of wheat, make gluten from each type. Form in ball. Bake on cookie sheet at 350°. The ball that rises highest has the most gluten content. Note: Gluten in Triticale wheat does not hold together, nor does it raise. Use other types of wheat for making gluten and bread. (Soft wheat indicates more moisture content and less protein.)

Milling The Wheat

There are many home wheat and grain mills on the market today which produce fine flour. Grocery stores usually carry whole wheat flour that is adequate for making gluten, although the yield may be lower.

Storing and Preserving Wheat

The best way to store wheat is in an airtight container, either plastic or metal. Hard wheat (low moisture content) can usually be stored without any chemical preservatives. If Fumigation is desired, one of the most effective and safest to use, in preventing weevil, is dry ice. Use 7 oz to 100 lbs (about 1/2 C in a 5 gallon can). Fill bottom of can with wheat. Place crushed ice in, then fill top with more wheat. Leave lid open about 30 minutes to allow fumes to escape. Seal. If any bulging occurs, remove lid and wait a few more minutes. Wheat markedly loses the capacity to germinate after 5 years of storage; however, the nutritive value is retained. Hard wheat, stored properly, will last many years. An estimated amount of wheat to store averages one pound per person per day (365 lbs. per person per year).
For your health's sake whole wheat should be introduced slowly into your system. For peace of mind, if you have stored wheat, you will know how to prepare it into tasty meals using these recipes.

Tools, Equipment and Bulk Buying

For information on these items as suggested in the book, if not available in your area, send a self addressed, stamped, legal sized envelope to LM Publications, Box 482 • Provo, UT 84603.

The beginning of a new era of cooking with wheat. Now for the first time this section brings a revolutionized method for separating gluten from the flour in only minutes. Also, what to do with the by-products left from the gluten separation? It is all here, everything from drinks, flakes, crackers, and pastry to ice cream!

SECTION 1

GLUTEN

GLUTEN (GLOOtun) is a mixture of proteins in wheat as well as some other cereal grains. It helps to make dough rise. Gluten in wheat flour is composed chiefly of two proteins, gliadin and glutenin. One aspect in utilizing the wheat grain to it's fullest is the isolation of this protein from the flour. This is done commercially by a vacuum process and is called Gluten Flour, or you can do it at home with your own ground wheat flour. Gluten can be separated from the other components by adding water and using a stirring or beating method, then a rinsing process. This nitrogenous tough, sticky, somewhat elastic substance is grayish in color, has almost no taste, is insoluble in water and can take on various textures when prepared with techniques found in this book.

The family cook, faced with the task of serving meat-eaters and abstainers at the same table, will find gluten an answer to the dilemma. When properly seasoned and prepared to resemble meat in both appearance and flavor, gluten makes converts of nearly everyone.

Gluten is a delicious vegetarian alternative to meat that is made from the protein of wheat. Meat of wheat creates the opportunity to turn traditional recipes into vegetarian ones.

Exact quantities are unpredictable in gluten making because of the varieties of flour. Use the guidelines in the book and you will have success with the enjoyment of wholesome eating.

Learning to prepare gluten makes a satisfying transition from meat to wheat.

ADVANTAGES OF USING GLUTEN

1- A Good Protein Food

Gluten is listed as a principal source of protein along with eggs, milk, cheese, lean meat, fish, soy beans, peanuts and vegetables. Gluten contains the 8 amino acids which make up protein; however, the amino acid lysine in most wheat is low. In order to make the protein in wheat complete at the same meal or at meals as much as two days apart, add to your gluten dishes any of the foods from the legumes and vegetable or nuts and seeds groups. Send for our Vegetarian Protein Planning Chart.

2- For Personal Health

Do your liver a favor and cut out meat

Heart disease, cancer, premature aging, and food poisoning have been linked to meat eating. Today's beef contains over 6 times more fat than it did in the 50's. Cutting down on meat definitely reduces high blood pressure and promotes a longer, healthier life. Chemicals and residues in concentrated amounts are found in animal meat. They include pesticides, fertilizers, hormones for weight gain and antibiotics to prevent disease. However, in routine use the bacteria becomes resistant and meat from diseased animals can cause food poisoning in humans. Tranquilizers, to make the animal content with confinement, and hormones have proven to be carcinogenic. These chemicals do not pass through the animal but remain in the meat, even after it is slaughtered.

Just before and during the agony of being improperly slaughtered, the biochemistry of the terrified animal undergoes profound changes as it futilely struggles for life. Toxic by-products and large quantities of adrenaline are forced throughout the body, thus pain-poisoning the entire carcass.

Eating less meat can help the starving.

If we ate half as much meat we would release enough food for the entire developing world. Only 10% of the protein and calories we feed to our livestock is recovered in the meat we eat. Seventy-eight percent of all our grain goes to animals. While the world is faced with mass starvation, we consume over a ton of grain per person per year (through our livestock and other goods,) while the rest of the world averages about 400 lbs. "The earth has enough for every man's need but not for every man's greed" - Mahatma Ghandi.

Meats are not easily digested by man,

his digestive system is not designed for meat eating. His bowels are 12 times the length of his body for slow digestion of vegetables and fruits, which are known for slow decay. Carnivorous animals have very short bowels, 3 times the length of the body, for rapid expulsion of putrefactive bacteria from decomposing flesh.

"We now know that protein increases calcium excretion more than it increases calcium absorption, thus leading to an overall loss of calcium from the body," writes Dr. Notelovitz.

Gluten is easily digested.
The tender texture of gluten makes it a superior substitute for meats. Using gluten in place of meat helps eliminate the build-up of uric acid and cholesterol factors present in meats.

3- Low Cost Meals

12 C of flour, prepares into 9 C ground gluten (286 grams protein) which is equivalent to using 3 lbs hamburger (288 grams protein). This makes gluten meals **one-fourth the cost of meat meals.**

12 C whole wheat flour makes 4 cups raw gluten
4 C raw gluten bakes into 9 cups ground gluten
9 C Cooked Ground Gluten Pieces is equivalent to 3 lbs hamburger (cooked)

By making your own "meat", you will have a greater variety of meals and save money as well.

For instance:
12 cups whole wheat flour (56¢) made into gluten yields:
512 gluten cubes-1/2 x 1/2 in. or
150 meat balls - average or
20 steak slices - 4 x 1/2 in. or
4 pie crusts - graham cracker type

4- Saves Time

It takes less time to make a batch of gluten for meals than to make a trip to the store for meat. Do several batches in an afternoon and save time. It only takes minutes more to double or triple a recipe. See planning meals ahead (page 64).

5- For a Variety in Food Preparations

Gluten can be made into just about any type of food you desire.
• **A Meat Substitute:** (no cholesterol)
snacks, casseroles, chicken, beef, fish, jerky, sandwich fillings, and salads. Use in place of hamburger in chili, spaghetti, pizza etc.

• As a Meat Extender:
Use only one-fourth ground meat, the rest Ground Gluten. Omit the beef, or other meat seasoning bases and prepare as the recipe indicates. note: This is a good way to introduce gluten to meat-lovers; however, most people cannot tell the difference when you leave the meat out altogether.

• As a Cold Cereal:
Use Gluten Crunch alone or include it in any granola recipe.

• For Topping and Desserts:
Use Gluten Crunch in puddings, gelatin salads, ice creams, apple crisp, cookies and cakes.

• To make Candies:
Use Gluten Crunch to make carob (or chocolate), candy, rice crispy treats, or as you would use nut meats in any of your favorite recipes.

APPROXIMATE COMPOSITION OF
COMMERCIAL GLUTEN FLOUR

Commercial Gluten Flour - 100 gm (grams) (454 grams = 1 lb)

Water	8.5%
Calories	378
Protein	41.4 gm
Fat	1.9 gm
Total Carbohydrates	47.2 gm
Fiber	0.4 gm
Ash	1.0 gm
Calcium	40 mg
Phosphorus	140 mg
Sodium	2 mg
Potassium	60 mg

Benefits of Gluten Flour

• Improves binding abilities
• Mimics texture and flavor of meat
• Can be stored in dehydrated form
• Enhances chewiness and texture
• Provides excellent source of protein

Storage of Gluten Flour

Gluten flour should be stored in cool, dry clean area free from odors. Optimum conditions 77°F and 50-60% relative humidity. Shelf life is 6 months when stored under optimum conditions.

MATERIALS AND EQUIPMENT
NEEDED FOR MAKING GLUTEN

Flour
Regular ground whole wheat flour is best, although white flour can be used. However, the gluten made from white flour is not as easy to control.

Water
Cool to lukewarm water is used in the rinsing process.

Seasonings
Gluten has no definite taste; therefore, a variety of flavors can be used. Many preparations and mixes are available at grocery stores, gourmet shops and restaurant supply outlets. See the section in this book on how to make your own seasoning mixes.

Tools
• Large bowl and spoon
• Colander — plastic or metal (not a wire strainer).
• Hand food grinder, food processor or blender for grinding cooked gluten
• Heavy baking sheet and a non-stick vegetable cooking spray
• Sauce pans, double boiler, vegetable steamer and/or pressure cooker

Note: Check the Question and Answer section at the end of this section for more assistance with Gluten making.

See the "Quick Wholesome Foods Video" where LeArta shows you close-up views of the gluten making process. See the exact texture and step by step instructions for successful gluten making.

The Quick Wholesome Video is available upon request. See info in back of book.

PRODUCING THE RAW GLUTEN

METHOD #1
RAW GLUTEN FROM COMMERCIAL GLUTEN FLOUR

Commercial Gluten Flour, (found in health stores or bulk buying outlets) produces instant gluten when water is added to it. To give this type of raw gluten a more tender texture, add any bean or whole grain flour (commercial or home ground) to the gluten flour before stirring in the liquid.

Mix together thoroughly
2 C Gluten Flour
1/3 C flour (soy, whole wheat, rice, etc.)

Combine
2 C hot water
3 T seasoning of your choice (soup bases, seasoning packets - onion soup, taco, etc.) or leave out to flavor later for recipes with specific flavoring suggestions.

For a darker color add 2 T Kitchen Bouquet (found in grocery stores)

Stir liquid into flours (takes only about 10 stirs) and you will have raw gluten. Because some Gluten Flours differ slightly you may have to add a sprinkle more water to barely moisten the flours.

METHOD #2
USING WHEAT FLOUR

This method is for those who have access to a flour mill and want to use their own homeground flour.

Stirring Technique (using a bowl and spoon)

Stir together (takes about 20 stirs)
12 C whole wheat flour
7 C water (or enough to moisten all the flour particles)
The dough should resemble a bread-like dough before being kneaded. Set this mixture aside for 20 minutes or longer.

The next step, (one you do not have to do when using Commercial Gluten

Flour) is the rinsing process.

Kneading Technique (using an electric bread mixer)

A good method to use if you have a low protein soft wheat

Add double the amount of flour to the amount of water you put in bowl (12 cups flour to 6 cups cool water.) Mix with kneading arm. The consistency should be like bread dough and pull away from the sides of the bowl. If it does not, add more flour. Mix 5-10 min. Now it is ready for the rinsing process.

RINSING PROCESS

To separate the gluten from other products in the wheat flour which has been stirred or kneaded.

Add a small amount of clear water to bowl of rested dough. Work and squeeze with your hands to loosen the dough (only a few seconds). When the water takes on a milky appearance and you see specks of bran, pour this water off, holding bulk of the dough back with your hands. Over a sink, place this dough in a colander (plastic best) with another bowl placed underneath the colander to catch any of the gluten that slips through the holes. Under a tap of slowly running lukewarm water, work and squeeze the dough with your hands until the gluten starts to hold together and the liquid coming from the dough is clear. It is not necessary to rinse out all of the bran from the gluten. In about 3 minutes or longer you should have a ball of elastic-like dough. This is the Raw Gluten.

Note: The dough becomes slightly stringy and falls apart easily just before it starts holding, so don't give up too soon, but if within 5 min. the gluten does not start holding together, it needed one of the following: More resting time, less water mixed with flour to make the dough, flour with higher protein content. Don't throw it away, try again! Add more flour to the, mixture and let rest again, only longer - 1-4 hrs., or even overnight.

Helpful Tips:
• Once a small amount of gluten starts holding together, you will find the rest of the gluten clings to it, so, as quickly as possible get a small ball of gluten started. You may want to work a small amount in your hands to get this start.

• How soon the gluten cells start holding together is determined by the protein quality of the wheat flour used, or how often the clear water is allowed to run through the dough.
• If your wheat is less than 15% protein, the gluten cells do not hold together as well and the raw gluten yield will be less. It will be helpful to either let the dough rest longer before rinsing it or to pour the milky liquid off more often.

PROCEDURE TO USE WHEN THE LIQUID, STARCH AND BRAN ARE TO BE SAVED

The liquid rinsed away, while isolating the gluten, contains vitamins, mineral-rich starch, bran and the wheat germ. It can easily be saved and added to recipes for that extra fiber and nutrition.

Follow instructions in the rinsing process. When you are ready to pour the milky water off, rather than pour it down the sink pour it into another bowl. Repeat this procedure of working the dough under running water until liquid is milky and ready to pour out again. Repeat until you have reserved as much liquid as you want to save. The first two pourings contain the most vitamin-rich liquid.

Starch and Gluten Water

After you have followed the rinsing procedure, pour the reserved liquid into a large glass container. It usually takes 1 gallon of water to rinse and separate the gluten when using 7-10 cups of wheat flour.

Let settle 2 hours or longer, or until you can see the definite layers of bran on the bottom, next starch and then water on top.
Slowly pour off the liquid on top into another container. Next, spoon off the starch, leaving the remaining bran at the bottom of the container.

Use these products in:
gravies, stews, desserts, soups, drinks, crackers, chips, pies, ice creams and pizza dough, crepes, breads, muffins, pastries, pancakes, waffles, and thickenings. Can also be used in water for plants, baby's bath, hand lotion, or any recipe calling for water, starch or bran.
Note: Refrigerate and use these products within 2-3 days, or freeze.

BRAN

Bran contains the very important trace minerals. It is a roughage that aids digestion, helps in colon disorders and is very important in the balance of the body's system. It is also high in phosphorus and potassium. Uses for the bran: cold cereals, crackers, drinks, cakes, cookies, pancakes, breads and many other recipes. Add a little to any batters, drinks or main dishes.

When you want to save only the bran, and not the water or starch:
At the beginning of rinsing process, place a deep pie plate or large bowl in sink under a plastic or metal colander (not a wire strainer). As the bowl or pie plate fills, or when you see the bran surfacing and spilling over, carefully pour off excess water and empty bran into another container. Repeat this saving process until most of bran is washed from gluten. If stream of water is too rapid, bran will be washed away and will not settle at the bottom of the container.

Straight Bran
(Without Starch Residue)
Place saved bran in clear glass container. Let rest 10-20 minutes. As white water comes to the top, pour off and replace with clear water. Repeat this process until water is clear at the top. Pour this water off and spoon the remaining bran into an airtight container. Bran will keep 4-6 days in fridge. It may be frozen.

Note: If the bran has pieces of gluten in it, pour bran into colander with plate or bowl underneath. Rinse and work pieces of gluten together with hands under running water, letting the bran settle again. Pour off excess water.
See Recipes using the saved bran, starch or water. (pg 59)

COOKING THE RAW GLUTEN

The next step after you have the raw gluten is to cook it. This produces a firm product which enables you to then prepare it into a variety of textures and shapes. The best method to cook the raw gluten, in preparing it to add to your meals, is a steaming method.

STEAMED METHOD

Place the raw gluten in any steaming device, sprayed first with a cooking spray. Any vegetable steamer, rice cooker etc. will work. Steam the raw gluten until it is firm, about 20 to 30 min.

For specific shapes

To produce a tighter texture (for chicken pieces, pepperoni and baloney-type slices) There are two methods:

#1 Form the raw gluten into a roll shape. Wrap cheese cloth around the gluten (overlapping a little) and secure it by tying both ends, and middle if necessary, with a string. Steam about 30 minutes

#2 Place raw gluten in No.2 1/2, or smaller, can (sprayed with a cooking spray) in the steamer. Steam 20 to 30 minutes. Turn can upside down and shake gluten out. Cut into slices or cubes.

SIMMERED METHOD

Used in making thin strips for stir-fry dishes, stroganoff, jerky etc. Roll out raw gluten on a slightly wetted plastic or Formica surface with rolling pin, about 1/4 inch thick. Cut with pizza cutter or knife, into strips.

Drop strips of raw gluten into a flavored boiling broth (use equal amounts of broth and raw gluten.) Simmer until liquid is gone, or about 30 minutes, stirring occasionally.

Dry out in food dehydrator or place pieces on cookie sheets that have been coated with a non-stick baking spray. Place in 350° oven (leave door ajar slightly) until pieces appear dry on top. Turn pieces over and continue to bake until texture is chewy, about 30-60 min.

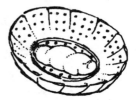

PREPARING GLUTEN INTO
SPECIFIC SHAPES

The following instructions should give you good results in using gluten in your meals both as a main dish and dessert.

Now that you have cooked the raw gluten it is ready to be prepared into various shapes and textures depending on the type of wheat meat dish you would like to substitute or variety of wholesome sweet treats.

GROUND GLUTEN
Grind the steamed or baked gluten in a food processor or hand food grinder on medium to large disc. This texture is used in making meat balls, patties and veggie burgers, sausage, meatloaf, as a meat extender or in any recipe calling for ground meat; including in casseroles, soups, salads, sauces, gravies desserts.

SLICED GLUTEN
By slicing it thick and flavoring it, you can have satisfying breaded cutlet-type steaks. Thinly sliced gluten is used in recipes and sandwiches calling for chipped beef, stir-fry strips, bologna, pepperoni, jerky, etc. See instructions for making Jerky strips on page 53.

CUBED GLUTEN - Cut into cubes of desired sizes and flavored with beef, chicken, ham, clam, crab, etc. to add to sauces, soups, stews, shish kabobs, sandwich fillings and salads.

FOR CHICKEN OR FISH-LIKE PIECES- With fork or hands break away from steamed gluten the size of pieces desired (steamed method for a tighter texture.)

SEASONING THE GLUTEN

If you have not already seasoned the raw gluten(as in the commercial gluten flour procedure) use the following instructions.

Seasoning Raw Gluten

When making gluten from commercial Gluten Flour, the flavor can be added BEFORE it is cooked. (see page 10- method #1)

Seasoning Cooked Gluten

It is hard to add seasoning into the raw gluten as it is a bubblegum-like texture and difficult to work in any type of flavoring. It will be added AFTER the raw gluten has been cooked.

Cooked ground gluten is flavored after it has been ground. Remember you cannot grind raw gluten. The recipes will give instructions on how to season and flavor.

For Gluten Slices, Cubes or Pieces - Place unflavored cooked gluten shapes in saucepan with enough seasoned broth of your choice to partially cover. Simmer 3-5 minutes, depending on thickness, turning if necessary.

Basic Broth:Mix 1 C liquid, 2 T flavored base, see section 2 for herb blends or use a soup base. (Chicken page 95, Sausage page 100)

SEASONINGS AND RECIPES FOR THE VARIOUS GLUTEN PREPARATIONS FOUND IN THIS SECTION

For those who prefer not to use animal products found in most soup bases, the Basic Sausage Seasoning on page 100 produces an all-around great flavor for just about any recipe calling for beef flavoring.

Ground Gluten pages 17-38
Gluten Steaks and Slices pages 39-42
Gluten Cubes & Pieces pages 43-45
Gluten Chicken pages 46-49
Gluten Strips pages 50-53
Gluten Seafood pages 54-56

Be sure to study the information on page 64 for preparing meals in advance. This can save you time and money.

GROUND GLUTEN
Used in making meatballs, patties, sausage, meatloaf,
meat extender, casseroles and in desserts.

Basic Rule: When using any cooked and prepared gluten pieces or meat balls in a recipe that calls for boiling or simmering ALWAYS add just before serving or to heat through.

Used as an extender: When using Ground Gluten as an extender with hamburger or other ground meats, it is not necessary to add beef seasoning. Remember, you will need only one-fourth the amount of meat; the rest gluten.

Leftover gluten:. Roast, steaks, cubes that become spongy, or any cooked gluten you don't know what to do with can be put through a meat grinder to make ground gluten.

RECIPES

Ground Gluten Pieces

A preparation of Gluten to take the place of
cooked ground meat

2 C ground gluten
2 eggs
2 T flour
2 T olive oil
2 tsp sausage, mock chicken, taco, pizza etc. seasoning,
   bouillon or base
2 tsp fresh or dried minced onion

In a mixing bowl, mix all ingredients except gluten. (I like to use an electric mixer.) Stir in ground gluten and mix well.

Spread out on cookie sheet coated with a non-stick vegetable spray.

Bake at 350° until firm, about 20 to 30 min. Let cool and break into pieces. Store in closed containers in fridge or freezer until ready to use.
Yield: 30 small 1 inch pieces.

Basic Meatball or Burger Recipe

2 C ground gluten
3 T finely minced onion or 1 T dry minced onion
1 T sausage seasoning, (pages, 100-101), chicken (page 95),
  or seasoning of your choice.
2 T flour
1-2 eggs, beaten
2 T oil (olive best)
salt and pepper to taste

To Form Meat Balls:
Mix ingredients together and form into balls. Bake at 350° on
cookie sheet 20-30 min. or until firm.

(Can store in fridge or freezer) Prepare for serving by cover-
ing with cream sauce or soups, sweet and sour sauce, tomato
sauce spiced with Italian or Spanish seasoning and let heat
through. Delicious in spaghetti, cheese or other casserole
dishes.

To Form Patties:
Mix ingredients together and form into pattie shape. Brown
in oiled skillet. If desired, patties may be dipped in egg or
breaded first before browning. Serve plain or topped with
gravy.

Variations:
Add to the basic Gluten Meat Ball or Burger recipe any of
the following:
• 1 C cooked brown rice
• 1/2 C chopped mushrooms (canned)
• 2 tsp Worcestershire sauce (for flavor and a darker color)

Once the balls are baked, serve with a variety of sauces and
gravies. (Try a Sweet and Sour, Spicy Tomato, Enchilada
Sauce or a Creamy Mushroom.)

*Hint: For a darker colored gluten ball, add 1 tsp Kitchen
Bouquet.*

Oriental Meat Balls *With ground gluten*

2 C ground gluten
2 T finely minced onion
1 T chopped green pepper
1 T beef seasoning, or base
1/4 tsp ground ginger
2 T flour
2 eggs
1 T soy sauce
1 T sesame seeds
1/2 tsp sesame oil
2 T oil (olive or canola)
salt and pepper to taste

Mix all ingredients except gluten together. (I like to use an electric mixer.) Add gluten and stir till moistened. Form in balls.

Bake on cookie sheet 30 minutes at 350°. OR, fry in heavy skillet coated with a non-stick vegetable spray.

Makes 30 one-inch balls.

Meat Balls for 50

20 C ground gluten
2 C chopped onions or 1/2 C instant onion flakes
1/2 C beef seasoning or base
1 1/4 C flour
20 eggs
2 C olive oil
2 T salt
1 T pepper
3 T parsley flakes
1 T sausage seasoning if you want slight sausage flavor

Combine all ingredients except gluten in large container. (I like to use an electric mixer.) Stir in gluten. Form into balls.

Bake on cookie sheets at 350° for about 20 min. Makes around 350 balls.

Cover meat balls with sauce and heat through 20 to 30 min before serving. A quick, tasty sauce, one that usually pleases most people, is cream of mushroom soup, thinned with 1/2 can milk.

Hint: When using part hamburger, replace 10 C ground gluten with 3 Lbs hamburger.

Quantity Cooking Information

1 lb raw diced potatoes = 3 cups
1 lb chopped onions = 2 cups
1 oz salt = 2 T(1 C = 8 oz)
1 oz pepper = 4 T (1 C = 4 oz)
1 oz nutmeg = 3 1/2 T
1 oz parsley = 1 bunch (15 sprays)
4 oz chopped parsley = 1 C (4 T = 1 oz)
2 medium green peppers = 1 C chopped (4 oz)

Gluten and Hamburger Meatballs with Cream Sauce

1 recipe "Gluten Meat Balls" uncooked (page 18)
1/2 lb hamburger
1 crushed clove garlic
2 tsp grated lemon rind
1 tsp celery salt
1 tsp Worcestershire sauce

Follow basic recipe. Form into balls, Bake or fry in vegetable oil until firm. Remove from skillet.

Sauce
1/2 C chopped green pepper
2 T flour
1 1/2 C tomato sauce
1/2 C water
1/4 tsp salt
1/2 C sour cream

Sauté peppers in skillet until tender. Stir in flour and brown (about 2 minutes). Add tomato sauce, water, salt and continue stirring. Add sour cream. Pour over meat balls and bake at 300° for 30 min. Serves 6.

Spicy Tomato Gluten "Meat" Balls

1 recipe gluten "Meat" balls (page 18)

Note: can replace 1 cup ground gluten with 1/2 lb hamburger Form into balls 1/2-inch in diameter. Bake or fry until browned on all sides. Set aside.

Sauce
2 T melted butter
3 T flour
2 C canned tomatoes
1/8 tsp pepper
1/2 tsp prepared mustard
1/2 C grated cheese
1/2 tsp salt
1/4 tsp Worcestershire sauce
1 T honey

Melt butter, stir in flour and add tomatoes. Stir until thickened. Add the pepper, mustard, cheese, salt, Worcestershire sauce and honey. Stir until cheese melts and pour over meat balls. Cover and simmer 20 min. Serves 6.

Zucchini Soup with Gluten Meat Balls

2 medium onion, chopped
2 T butter
1 1/2 Lb zucchini, sliced
3-4 C chicken broth
1/8 tsp black pepper
1/8 tsp nutmeg
1/8 tsp salt dash cayenne
2 C Gluten "Meat Balls
1/2 C cream or yogurt
Parmesan cheese

Sauté onion and butter. Add to zucchini and broth. Simmer for 5 min. Add spices and blend well in blender. Pour blended mixture in saucepan with gluten balls; heat through. Add cream. Sprinkle Parmesan on top and serve. Serves 5 - 6

Seasoning Variations to Basic Burger recipe

Barbecue Burgers -   Add 1 tsp Spike or Season-ALL, 1 tsp barbecue spice or 1 T BBQ sauce. Omit salt.

Mustard Burgers -   Add 1/2 tsp dry mustard, 2 T catsup.

Herb Burgers -   Add 1/8 tsp each of thyme, marjoram, basil; 1/2 tsp celery salt; 1/2 tsp dill seed and 1 tsp parsley flakes.

Oriental Burgers -   Add 1/4 tsp ginger, 1 tsp lemon peel, and soy sauce to taste.

Pepper Burgers -   Add 1 1/2 T green or red sweet bell peppers, chopped and 1/4 tsp savory salt.

Veggie Burger

Mix together :

4 eggs
3 T olive oil
2 tsp sausage seasoning
1 tsp each garlic and onion powder
1/2 tsp salt
1/4 tsp pepper
1/2 C potato or oatmeal flakes
1/4 C flour (whole grain or bean)

Add to :

4 C ground gluten
1/2 C grated vegetables (about 4 T each green and red bell peppers, celery, onion and carrots) Note: could use dried vegetables, softened.

Spoon onto preheated, medium heat fry pan and brown on both sides or place formed patty on a baking sheet & bake 20 minutes or until firm. Serve on bun with fry sauce, slice of cheese, thin slice of red onion, pickles, lettuce, sprouts or your favorite condiments. Makes 14 patties (1/4 C each). Can freeze by layering with wax paper and placing in airtight container or zip-lock baggie.

Basic Gluten Sausage Recipe

Used for replacing meat sausage in pizza, chili, tacos, casseroles, pasta and vegetable dishes.

2 C Ground Gluten
2 eggs
2 T flour
2 T Olive Oil
1 T Basic Sausage Seasoning Mix (page 100)

In bowl mix all ingredients except gluten (I use an electric hand mixer.) Add gluten and stir well. Form into patties, if desired. Bake at 350° for 20 min. to develop a solid texture. This makes it easier to fry as well as easier to store until ready for adding to recipes. Serve hot.
Note: If mixture seems dry add more oil.
For a pepperoni flavor: add 2 tsp powdered anise seed and 1 tsp fennel seed to mixture above.

A Sausage Alternative
After ingredients are combined, spread out on cookie sheet. Bake 20-30 min. at 350° or until firm but not crisp. Remove from oven and crumble to resemble cooked ground sausage. Store in freezer or refrigerate.

Use in recipes calling for ground sausage or hamburger.

Serve in patty form for breakfast with eggs and pancakes. Add sausage seasoned gluten to meat loaves, poultry dressing, patties or use as an extender for other meats.

Meatloaf With Gluten and Hamburger

3 C ground gluten
1 lb hamburger
2 eggs, well beaten
6 T finely chopped onions
4 tsp meat loaf seasoning mix
salt and pepper to taste

Mix all ingredients and bake 1 hr at 350°.
Variation:
Add to mixture 1 can cream of mushroom soup, 8 T ketchup
or 1 T horseradish.

Savory Vegetarian Meatloaf

1 C grated raw potato
1/2 C grated onion
1 C chopped celery
1/2 C rolled oats
1/3 C oil
3 C ground gluten
4 eggs, beaten
1 tsp leaf sage, crumbled
1 tsp meat loaf seasoning (page 94)
1 tsp soy sauce
1/2 tsp celery salt
1/2 tsp salt

Mix all above and place in oiled baking dish. Bake at 375°
one hr. Slice and serve with sweet and sour sauce or gravy.

Sausage Filled Mushrooms

A gourmet hors d'oeuvre

Rinse mushrooms off slightly with water. Dry with paper
towel and scoop out stem, plus part of mushroom to make a
small well. Place in well of each mushroom 1/2 teaspoon
melted butter or olive oil, then fill generously with Basic
Sausage recipe (page 24,). Place in baking dish and bake 15
min at 350°. Serve as an appetizer.

Stuffed Peppers

4 lg green bell peppers
3 C Cooked Ground Gluten Pieces
1 onion, chopped
2 garlic cloves, minced
1 can tomato sauce (8 oz)
salt and pepper
2 eggs, beaten
flour (about 1 C)

Broil peppers until skins turn brown. Peel skins off, remove core and seeds. Sauté onion, garlic. Add gluten, salt and pepper, and tomato sauce. Stuff the mixture into peppers and squeeze closed. Roll in flour, next in beaten egg, then in flour again. A batter could be used in place of flour and egg. Fry in deep oil. Drain on paper towel.

Green Pepper Bake

Mix:
2 C Cooked Ground Gluten Pieces (page 17)
2 tsp Italian seasoning (page 93)
1/2 tsp salt
1/4 tsp pepper
2 eggs
1 C rice, cooked

Cut the top off or halve 4 large green bell peppers and remove seeds. Parboil if desired. Place stem-side down or pepper half in baking pan and fill with above mixture, replace top, pour a broth over this, about half way up. Cover and bake in 350° oven, 30 min.

Stuffed Cabbage Italiano

2 C Cooked Ground Gluten Pieces
1/4 C chopped onion
1/4 tsp thyme
1/4 tsp ground black pepper
1/2 tsp salt

1 tsp Italian seasoning
1 C rice
2 eggs, beaten
12 cabbage leaves, parboiled
1 8 oz can tomato sauce
1 T brown sugar
1 T lemon juice

Combine gluten, onion, seasonings, rice and egg and mix
thoroughly. Place about 3 T mixture on each cabbage leaf.
Fold sides of leaf in and roll tightly, starting from the thickest
end. Secure with toothpick or tie with string. Place in skillet.
In bowl, combine tomato sauce, brown sugar, and lemon
juice. Pour over cabbage rolls. Simmer covered, basting
about 3 times for 1/2 hr or till done. Serves 6.

Spanish Rice with Herbs

1/2 C green bell pepper, chopped
1/2 C onion, chopped
1/4 C oil or fat
1 1/2 C water
1/2 C Picante sauce or salsa
2 C can tomatoes or 2 C tomato sauce
1 T Sausage Seasoning
3/4 C uncooked brown rice
1 tsp brown sugar
1/2 tsp Worcestershire sauce
Dash of cumin, oregano, salt and pepper to taste
1 C Cooked Ground Gluten Pieces or Cooked Gluten strips

Sauté green pepper and onion in oil or bacon fat until tender
but not brown. Add remaining ingredients, except the gluten.
Cover and simmer 45 min. Add gluten pieces and heat thoroughly. (When using precooked rice, simmer uncovered 5
min or till rice is tender). Crumble bacon bits on top. Trim
with parsley. Serves 6

Picadillo Empanadas

A delicious meat or fruit-filled turnover

Pastry for a 2 crust pie (page 184)
1 egg, beaten

Filling:
3 C Cooked Ground Gluten Pieces
1/2 C chopped onion
1 garlic clove, minced
1 C tomato sauce
1/2 C green olives, chopped
1-2 tsp chili powder mix

Simmer onion, garlic, tomato sauce and beef seasoning until thick. Add gluten, olives and mix together.

Place a spoonful of mixture on 2 1/2 in x 2 1/2 in rolled out round or square pastry pieces. Dot edges with either water or beaten egg before placing top over mixture and bottom crust. Seal edges with fork or hand and brush top with egg for a shiny, brown look.

Bake 15 - 20 min at 350°. Make ahead, freeze and cook to serve.

Sour Cream Enchilada

5 corn tortillas
1 C Cooked Ground Gluten Pieces
1 C beans, refried (see "Rita's Bean Book")
2 1/2 C shredded cheese, reserve 1 C for topping
1/4 C sliced ripe olives
1 green onion, chopped
2 C sour cream, reserve 1 C for topping

Combine gluten, beans, cheese and 1/4 C of the sauce.
Spoon 1/4 C of the sauce in a casserole dish. Soften tortillas one at a time by dipping them in remaining sauce.

Place gluten-cheese mixture down center of each tortilla. Next spread about 1/4 C sour cream in each then roll up and place seam side down in baking dish.

Spread remaining sauce and sour cream then sprinkle remaining cheese, olives and green onions on top. Bake for 20 min at 350° or until hot and bubbly.

Basic Enchilada Sauce

2 C tomato sauce (two 8 oz cans)
2 T roasted and peeled green chilies
Add
1/2 tsp garlic powder
1/4 tsp oregano powder
1/2 tsp cumin (cominos)
1/2 tsp powdered bay leaf (laurel leaf)

In sauce pan simmer, 10 min.

Variations: To basic sauce add 1 T peanut butter
Try equal parts mayonnaise or salad dressing and the Enchilada Sauce for a dressing on fresh salads or for a dip (similar to French dressing).

Garlic Enchilada Sauce

2 C tomato sauce
1/3 C water
2 tsp chili powder
1/8 tsp cumin
1/4 tsp turmeric
1 garlic clove, minced
dash of cayenne pepper

Simmer 20 min

Tacos *Use the following recipes for making tacos*

Roasted Green Chili Sauce

2 C tomato sauce (two 8 oz cans)
1/2 tsp olive oil
1 T whole, roasted, peeled green chilies
2 T grated fresh onion or 1 tsp instant onion flakes
1/2 tsp cumin
1/2 tsp salt
2 C Cooked Ground Gluten Pieces

Mix ingredients. Simmer 5 min.

Serve on corn tortillas with grated cheese, onion, lettuce, or alfalfa sprouts.

There is a Difference!
Whole, roasted, peeled green chilies (hot)
Whole, roasted, peeled jalapenos green chilies (extremely hot)

Variations: Place sauce on heated corn or flour tortillas that have been fried in oil until brown. Stack on top of this, any or all of the following:

Cheese, grated
Red onion, chopped; or fresh green onions
Red beans, cooked
Garbanzo beans, cooked
Cucumber, chopped
Green bell peppers, chopped
Red beets, fresh or canned, diced
Green peas, fresh or frozen (thawed, not cooked)
Lettuce and/or alfalfa sprouts

Top with a dressing of your choice, such as Buttermilk, Ranch or Italian Dressing

Chili Verde

2 C red beans (51/2 C cooked)
1 chopped onion
4 T chopped green pepper
3-4 T green chiles
1 garlic clove, crushed
2 T Worcestershire sauce
1 T Kitchen Bouquet
2 C tomato sauce
1 tsp salt
1/4 tsp pepper
1 C Cooked Ground Gluten Pieces
2 T olive oil
1 tsp prepared mustard
2 tsp cumin
2 tsp chili seasoning mix

Rinse off beans; cover with 4 C water and bring to a boil for 2 minutes. Remove from heat, cover pan and let sit 1 hour. Drain. Add 3 C more water and cook until beans are just barely tender (about 30 min).

Add:
Sautéed onions and green pepper , chiles and combine with garlic, Worcestershire sauce, tomato sauce, Kitchen Bouquet, salt and pepper.

Simmer 30 min. Add: Cooked Gluten Pieces and oil and season to taste with about 1 T molasses and 2 T sugar or honey. Simmer another 5 min.

Serves 6

Chili Pronto

Stir into a no. 2 can red kidney beans:
1/2 C water
1 can condensed tomato soup
1 1/2 tsp chili seasoning mix
Add:
1 T Worcestershire sauce
1 tsp Kitchen Bouquet and Cooked Ground Gluten Pieces

Heat and serve.

Chili is even more flavorful after the first 24 hrs. It may be made in advance and reheated.

Pizza Supreme

Crust
1 C water
1 T yeast
1 tsp honey
1 tsp salt
2 T oil (for a more chewy crust use less oil)
3 1/2 C flour (could mix 1/2 white)
Makes two 12 in. pies

Dissolve water, yeast and honey together. Add remaining ingredients. Mix and knead together. Let rise once. Divide dough into two parts. Roll 1/4" thick and place on pans. Brush dough with olive or salad oil.

Bake crust at 350° for 8-10 minutes (This is a preliminary baking). Take out of oven and cool.

Spread sauce on top of crust. Sprinkle on 3-4 C grated Italian or other white cheese such as mozzarella.
Top with 1/2 to 1 C each of any of the following: green peppers, onions, mushrooms, anchovies, olives etc. Sprinkle with 2 C. Cooked Ground Gluten Sausage Mix.

Place in 375° oven until cheese melts, about 8-10 min.

Pizza Sauce *True Italian Flavor*

> 4 T grated Romano cheese
> 1/2 tsp garlic powder
> 1/2 tsp oregano powder
> 1/4 tsp black pepper
> 1/2 tsp salt
> 1/8 tsp paprika
> 1/8 tsp nutmeg
>
> ***Add to:***
> 2 C tomato sauce (15 oz can)
>
> Mix all ingredients. Simmer 15 min. Covers two 12 in. pizza crusts.

Speedy Sauce Variations - cook as above
#1 • 2 cups tomato juice
• 3 T flour
• 1 T Worcestershire sauce
• Italian seasoning

#2 • 2 C V-8 juice
• 3 T flour
• Italian seasoning

Spaghetti

> **Sauce:**
> 2 garlic cloves, minced
> 1 large onion, finely chopped
> 2 C tomato sauce (15 oz can)
> 1 1/2 C tomato paste (12 oz can)
> 2 T Worcestershire sauce
> 2 C stewed tomatoes
> 3 C water
> 3 tsp Italian seasoning
> 1/2 tsp chili powder mix
> 1/2 C Romano grated cheese (in a jar or can)
> 2 T olive oil
> 2 C Cooked Ground Gluten Pieces
> *(continued)*

Mix all ingredients except olive oil, cheese and gluten; cover. Reduce heat; simmer 30-40 min, stirring occasionally.

Uncover; add remaining ingredients, simmer 15 min. Place steaming hot in center of warm plate of hot boiled spaghetti. Accompany with garlic bread. Serves 12

Sauce Variation:
1 tsp Italian seasoning
1/2 tsp garlic powder or crushed garlic clove
1/2 tsp chili powder
8 oz can tomato sauce (1 C)
1 #303 can tomatoes (2 C)

Mix all ingredients. Simmer 30 minutes. Add Ground Gluten Sausage Mix and simmer another 15 min.

For a good addition see Garlic Bread Spread (page 124)

Toasted Garlic Butter French Bread

1 cube melted butter (1/2 C)
4 to 5 cloves minced garlic

Mix together. Cut French bread in thick slices. Spread butter mixture on bread with brush. Sprinkle with Italian Seasoning (optional). Broil until browned.

Lasagna

4 C Cooked Ground Gluten Pieces
3 T cooking or olive oil
1 crushed garlic clove
2 T Italian seasoning (page 93)
1 T whole basil
1 tsp salt
1/4 ground black pepper
three 8 oz cans tomato sauce or 2 cans sauce and
    1 can tomatoes
3 T Parmesan cheese or grated Romano
1-1 1/2 C water
12 lasagna noodles
3 C cottage cheese or Ricotta cheese
2 T parsley flakes
2 1/2 C mozzarella cheese, slice or grated

Heat gluten, garlic and oil together. Stir in seasonings (1 envelope spaghetti sauce mix could be used in place of other seasonings), tomato sauce, Parmesan cheese, and water (to desired consistency). Simmer 10 min.

Cook noodles according to package directions.
Spoon a third of the sauce into baking dish. Lay one half of the noodles on top. Add one half of the cottage cheese and half of the parsley and a third of the mozzarella cheese. Add another 1/3 of the sauce and repeat layers. Top with sauce and then the grated mozzarella cheese.

Bake at 350° for 30 min or until cheese melts and sauce is bubbly. Let stand 10 min before cutting in squares and serving. Serves 12
Can be refrigerated until ready to bake (allow 15 min longer cooking time).

Quick Lasagna

12 lasagna noodles, cooked

Sauce:
2 cans (15 oz.) spaghetti sauce
2 cans (15 oz.) tomato sauce
2 1/2-3 C Cooked Ground Gluten pieces
2 C mozzarella cheese, sliced or grated
1 carton (small) cottage cheese
Parmesan cheese to sprinkle on each layer.

Layer noodles, sauce and cheeses, noodles, sauce, and
cheeses. Bake covered at 350° for 20-30 min. or until bubbly.

Sloppy Joes "Manwich Sandwich"

1/4 Lb hamburger (optional) cooked
1/4 C chopped onion
1/4 C green bell pepper, chopped
1 tsp salt
1/4 tsp pepper
1 T oil
3 C Cooked Ground Gluten Pieces

Sauté vegetables in oil. Add remaining ingredients and heat
through. Add Sauce.

Sauce:
3/4 C catsup
1/2 C water
1 T honey or brown sugar
1 T vinegar
1 T Worcestershire sauce

Simmer 30 min
Serve with cheese.
Makes enough for 18 buns

Ham 'n Dill

1 C Ground Gluten
1/4 tsp sausage or ham flavor
Mayonnaise (thinned slightly with pickle juice)
diced gherkins
chopped green onions

Mix all ingredients. Spread on bread. Include lettuce and you
have a very tasty cold sandwich. Add any other ingredients
you like.
Note: Create your own fillings. Here are some suggested
ingredients: Sliced cucumbers, minced parsley, chopped
watercress, avocados, fried green tomatoes, anchovies, chili
sauce mixed with blue cheese, cream cheese with chopped
dates on date bread, sautéed mushrooms and mayonnaise,
sliced radishes, sliced tomatoes. Mix tuna, or the liquid from
tuna, with gluten.

SWEET TREATS MADE WITH GROUND GLUTEN

Gluten Crunch

*Gluten Crunch can be used in many gluten
sweet treats.*

2 C ground gluten
1 C coconut, unsweetened or sweetened
3 T melted butter or margarine
4 T honey or raw sugar

Combine ground gluten and coconut. Melt butter and honey
together. Mix all together. Spread on cookie sheet and bake
25-40 min at 350-375°. (Cooking time and temperature
depends on how much of the mixture you put on a sheet.) Stir
occasionally in order to roast more evenly. Watch carefully so
it won't burn. The mixture will take on a dry and roasted
effect when done. Let cool. The end product should be crispy.
(continued)

Use In:
Topping for ice cream, puddings and on cakes before baking for a crunch self-topping; pie crust, cold cereal, candies, and substitute for nutmeats.

Hint:
Make 3 or 4 times the recipe, store in an airtight container, either on shelf or in freezer, thus you will always have some on hand for those last minute needed treats or toppings.

Quantity Amount:
8 C ground gluten
4 C coconut
1 C butter
1 C sugar

Yield: 2 1/2 -3 qts.

Recipes using Gluten Crunch

GLUTEN STEAKS AND SLICES

Imagine preparing mouth-watering Chicken Fried Steaks, Salisbury steak and Veal Parmesan - all with no cholesterol! Follow the basic recipe for gluten steaks (see pages 14-16). Use any of your favorite steak recipes.

Basic Gluten Steak Recipe

6-8 steak slices, simmered in chicken or beef broth
2 eggs
1 C breading: soda crackers, seasoned flour or bread crumbs

Dip steak slices in beaten egg, then into breading. Brown in skillet with oil and cover with gravy or sauce.

Seasoned Flour

Basic
3 T flour
1/4 tsp each, onion or garlic powder,
salt and pepper to taste
Mix together.

Zesty
Mix together:
2 C whole wheat flour
2 T salt
1 T celery salt
1 T pepper
2 T dry mustard
3 T Paprika
1/2 tsp ginger
1/2 tsp thyme
1/2 tsp sweet basil

Mild
Mix together:
1 C whole wheat flour
2 T powdered milk
1/4 tsp paprika
(continued)

1/4 tsp garlic powder
1/4 tsp onion salt
1/4 tsp ground black pepper
1 tsp salt
1 T chicken seasoning

Basic Crunchy Breading

1 part flour
2 parts crushed soda crackers, flakes or bread crumbs

Mix ingredients together and dredge moistened food in it for frying.

Parmesan Breading

1 part Parmesan grated cheese
2 parts crushed dry bread crumbs or flakes
1/4 part flour

(These recipes can be doubled and stored in the refrigerator or freezer)

Swiss (Salisbury) Steak

8 gluten steak slices
3 eggs
4 T seasoned flour
2 C Swiss steak sauce

Dip steaks, simmered in beef broth to flavor, into beaten egg then in seasoned flour. Fry in oil (olive or canola best) till browned on both sides.

Pour Quick Steak Sauce or Chunky Tomato Steak Sauce over steaks (should almost cover them, add water if necessary). Cover pan and simmer to heat through

Quick Steak Sauce

3 C V-8 juice
3 T finely chopped onions or 1 T dried minced onion
1 T Italian seasoning
3 T chopped green peppers

Mix. Simmer 5 min and pour over steaks.

Chunky Tomato Steak Sauce

2 C V-8 juice
1 tsp chili powder
1/2 tsp sweet pepper flakes
1/2 tsp parsley flakes
1 crushed fresh garlic clove
    or 1/2 tsp garlic powder
2 tsp raw or brown sugar (optional)
1 tsp salt
1/4 tsp black pepper
1/4 tsp ground oregano
1/4 tsp ground basil
1/4 tsp ground rosemary leaves
1/8 tsp nutmeg
1/8 tsp ground fennel seeds
1/8 tsp powdered allspice
1 onion, chopped
2 C. sliced fresh or canned mushrooms.
3 T green pepper (optional)

Mix all ingredients. Simmer 30 min. This sauce is even better after it has been stored in the refrigerator overnight.

Note: For a tasty spaghetti sauce, replace some of the juice with tomato paste.

Veal Parmesan Steak

6 gluten flavored steak slices, ready for dredging
2 eggs
1/4 tsp paprika
salt and pepper to taste
1/2-3/4 C Parmesan cheese

Add paprika, salt and pepper to eggs. Beat well.

Dredge gluten steaks in egg mixture, then in Parmesan cheese.

Fry in butter or oil until browned. Serve hot.

Variations:
• For that extra cheese flavor, sprinkle grated mozzarella cheese on top of each steak after it has been browned.

• Add to the Parmesan cheese, 1/2 cup bread
• crumbs and 1 T parsley flakes; brown and simmer in Italian seasoned tomato sauce.

Serve with creamed vegetables and soup.

Light French Sauce

Combine and cook until thick:
1 C liquid (water, broth, or gluten water)
2 T Cornstarch
1 T parsley, snipped into small pieces
1 T green onion, chopped

Add:
2 t lemon juice
1 C halved green grapes

STEWS AND SOUPS

Gluten Cubes

See pages 14 and 15 for preparing Gluten Cubes to add to recipes.

Chunky Stew

2 C gluten cubes
4 C brown gravy
1 bay leaf (remove before serving)
1/2 tsp parsley leaves
vegetables (potatoes, carrots, celery, onions)

For a more spicy flavor, add:
pinch of saffron
1/8 tsp lovage, crushed and dried (or use 1 T fresh)
1/8 tsp paprika

Cook vegetables and spices in water. Use this water for the liquid in making gravy.
Add vegetables and firm gluten cubes (cubes should be very firm and chewy) to the stew mix 15 min. before serving.
Use this basic recipe for the base of your favorite stew.

Spanish Stew

Sauté in olive or vegetable oil:
4 green onions or 1 white onion, chopped
1 clove garlic, minced
1 green pepper, chopped

Cook together until potatoes are tender:
2 C cubed potatoes
4 C canned tomatoes

(continued)

Combine both cooked mixtures and add:
2 T chopped parsley
1/2 tsp dried basil or 1 tsp fresh, chopped
1 tsp cumin and chili powder to taste (approx. 2-3 tsp)
1 tsp salt and pepper to taste
2 C firm gluten cubes

Simmer all ingredients together until heated through. Thicken the stew if necessary with bean flour or wheat flour-and-water paste.

Hearty Vegetable Soup

1 C Cooked Ground Gluten Pieces or Cubes
2 C canned tomatoes
2 C cooked whole wheat, beans or barley
4 C water
2 T olive oil
1 medium onion
1 C zucchini with peel, sliced
1 C carrots, sliced
1/2 C cubed potatoes (optional)
2 C diced celery
1 bay leaf (remove before serving)
1 T parsley flakes
1/2 tsp garlic salt or 1/4 tsp garlic powder
2 tsp salt
1/2 tsp pepper
2 T beef seasoning

Combine all ingredients except Gluten. Cover and simmer 30 min. Add the Gluten pieces 15 min before serving or till just heated through. Serves 7.

Two-Bean Vegetable Soup

Combine and simmer 20 min:
1 C diced carrots
1 C diced potatoes
2 C chopped canned green beans
2 C canned garbanzo beans
1/2 C chopped red sweet pepper
1 crushed clove garlic
1 tsp salt
1/2 tsp pepper
5 C vegetable, beef or chicken broth

Add 1 C Cooked Ground Gluten Pieces , or cubes. Simmer
20 more min.
Serves 4.

Chilled Summer Soup (Gazpacho)

*Very refreshing on a hot day. Excellent
served in small amounts for a cocktail.*
1 C zucchini, sliced in bite sizes lengthwise, not peeled
1 C onion, green or red, chopped
1 C green bell pepper, chopped
4 C V-8 juice
2 C fresh tomatoes, chopped
1 C Cooked Gluten pieces or firm gluten cubes (1/4"-1/4")
Seasoning to taste - pinch of the following is suggested: cumin,
  paprika, celery salt, Spike or Vege-sal, salt and pepper.

Sauté vegetables slightly and combine with juice in sauce
pan. Add tomatoes and Gluten meat balls or Cooked Ground
Gluten Pieces. Bring to boil and remove from heat.

Chill. Top with Parmesan cheese if desired before serving.

CHICKEN SEASONED GLUTEN

GLUTEN MOCK CHICKEN

You will find the mild herb flavor developed for gluten chicken dishes (page 95) to be very versatile. Use the flavored gluten chicken in any of your favorite chicken recipes.

Preparing:
To make gluten chicken pieces or steaks follow cooking instructions for raw gluten on page14, under Steamed Gluten. Remove the cheese cloth from the roll of cooked gluten and cut into desired shapes. Note: The chicken seasoning can be added to the water in the making of raw gluten from commercial gluten flour. When making raw gluten from your regular wheat flour you will need to season it by simmering the cut chicken-like pieces in a broth of 2 T Mock Chicken Seasoning Mix (page 95) to 2 cups water for 5-10 minutes. Pat dry and freeze what you do not use.

RECIPES FOR CHICKEN SEASONED GLUTEN

Chicken Broccoli Casserole

> 2 pkg frozen broccoli spears or asparagus (or the fresh equiv-
> alent) steamed slightly (about 3 cups)
> 3 cans cream of chicken soup
> 1 C mayonnaise
> 2 T lemon or lime juice
> 1/2 tsp curry powder
> 1-2 C chicken seasoned gluten strips (about 3" x 2" x 1/4"thick)
> paprika
> Parmesan cheese

> Place broccoli on bottom of 9 x 12 pan or glass baking dish.
> Arrange gluten chicken pieces on top of broccoli. Combine
> mayonnaise, cream of chicken soup, lemon juice and curry
> powder. Spread over ingredients in pan.

Sprinkle with Parmesan cheese and paprika. Bake 30 min at 350° or until top is bubbly and brown. Serves 8.

Serve with baked potato.

Note: Can be prepared the day before and refrigerated. Add Parmesan cheese and paprika just before baking. For a more tender broccoli (when using fresh), steam slightly before placing in pan.

Additional Gourmet Ideas

Chicken Florentine

Breaded, browned Gluten chicken steak served with cream sauce on a bed of spinach.

Chicken Cacciatore or Marinara

Breaded, browned Gluten steak heated in a chunky tomato Spaghetti sauce. Green peppers and oregano added for the Cacciatore. Serve with rice or pasta.
Include in sauce, sauteed green peppers and a pinch of oregano.

Chicken a la King

Chicken flavored gluten pieces added to a cream sauce with pimento and cooked green peppers added. Serve over patty shell, rice, toast or toast cups (a slice of bread buttered on both sides pressed into a muffin tin and Baked at 375° for 10 minutes.

Herb Poultry Stuffing

1/2 C flour
1/2 tsp salt and pepper, or to taste
3 eggs, beaten
2 C chopped nut meats (pecans, pine, walnuts, or brazil)
1/2 to 1 C chicken broth to moisten (see page 16)
4 T butter, melted
3/4 C chopped celery
1/2 C fresh minced onions or 2 T instant onion flakes
1 C canned or fresh chopped mushrooms
6 C Ground Gluten

Mix dry ingredients together. Add beaten egg.
Add liquid and remaining ingredients. Toss well
Makes 6-7 C of dressing Enough to stuff a 12-14 lb
bird. If baking separately, put in baking pan and bake 30 min
at 350°.

Variation:
Cut recipe in half and use to stuff pork chops.

Hint: Substitute 2 C ground gluten with 2 C dry bread cubes
for a lighter texture.

Baked Chicken Casserole

2 C uncooked poultry dressing
1 C thin gluten chicken pieces or cubes
3 C Chicken gravy
buttered crumbs

Place a layer of poultry dressing in bottom of baking dish,
then a layer of gluten and one half of the gravy. Repeat.
Sprinkle with buttered crumbs. Bake 30 min at 350° Serves 6

Variation: For an oriental flavor add to the above recipe 1/2
C. slivered almonds or water chestnuts, 1 T soy sauce and 1/2
T flavored sesame oil (found in the oriental section of your
supermarket).Top with crispy Chinese Noodles.

Basic Quiche Custard

1 1/2 C heavy cream
3 large eggs
salt, pepper, grated nutmeg to taste

Vegetable Quiche

Into a baked pastry shell place the following:
1-2 C vegetable of your choice - steamed to barely tender,
    (spinach, broccoli, mixed frozen vegetables etc.)
1/2 - 1 C fresh mushrooms, sautéed in vegetable oil
2 T onions or scallions also sautéed
1/2 - 1 C grated Swiss cheese

Over this pour the custard to which has been added 1 tsp
fresh or bottled lemon juice and 1/8 tsp nutmeg.
Sprinkle 3 to 4 T grated Parmesan cheese over the filling and
dot with 1 T butter. Bake at 350° about 30 minutes, or until
custard is set.

Quiche with Gluten Chicken

9 in. unbaked pastry shell
1 C gluten chicken pieces
1/2 c. chopped onions
2 T butter
2 T chicken seasoning or base
1 1/2 C milk, heavy cream or part yogurt
3 eggs
1/4 C shredded jack or cheddar cheese

Cook onion in butter until tender. Beat together chicken seasoning, milk and eggs. Combine all ingredients. Pour into 10"
pastry shell or quiche pan and sprinkle with grated cheese.
Bake for 25-30 min at 375° or until knife inserted in center
comes out clean.

STRIPS FOR STIR FRY

Follow instructions for making strips from raw gluten (page 14) using a broth seasoned with beef or chicken soup base.

Oriental Gluten Strips

2 C steamed gluten, which has been cut into strips of about
  1" x 1/2" and 1/8" - 1/4" thick.
2 C broth (2 T seasoning base added to liquid)
1 T soy sauce
1/2 t ginger

Simmer, stirring occasionally until liquid is gone. For a more chewy texture, dry out in a 350° oven for 20 min or longer. Leave oven door ajar. Can eliminate simmering in broth and marinate instead. (suggested marinade: 1/2 C Oriental Sesame oil, 2 cloves fresh crushed garlic, 1 T Soy Sauce.) See also page 14

Stir Fry

In a wok or heavy skillet sauté vegetables until tender. Do one type at a time, add to large pan. Cover until all are done.

1 carrot, cut diagonally
1 C cauliflower, sliced
1 C broccoli, flowers and peeled stems, sliced
1 green pepper, sliced
1 C celery, chopped
1 onion, sliced
3-4 green onions, sliced
1 garlic clove, minced
1/4" piece of fresh minced ginger or 1/4 ginger powder
2 C bean sprouts
1 C sliced mushrooms (optional)
toasted cashews or sliced almonds (optional)
Add: 1-2 C Oriental Gluten Strips and stir in 1 - 2 tsp sesame
  oil (found in oriental section of grocery stores.)

Reheat altogether. Serve over rice and top with soy sauce or chow mein sauce.

Chow Mein Sauce

1 C beef or chicken stock (or 2 T soup base to 1 C liquid)
2 T honey or brown sugar
1/4 C soy sauce
1/8 tsp dry mustard
2 1/2 T corn starch

Mix following ingredients in sauce pan and heat until thickened.

Sweet and Sour Sauce

1 C pineapple juice
1/4 C white wine vinegar
1 T soy sauce
1/3 C brown sugar or honey
1/4 C catsup
2 T cornstarch or Thick Jel (page 215)
Mix together in saucepan and cook over medium heat until
thick, stirring constantly.

Hawaiian Sweet and Sour *Using Gluten Strips*

2 T butter or margarine
1/2 tsp curry powder
1 C diced celery
1 chopped medium onion (1/2 C)
1 green pepper cut in thin strips
1 T corn starch or gluten starch
2 T brown sugar
1 tsp vinegar
1 tsp soy sauce
1 T Worcestershire sauce
1/2 C orange juice
1 can (9 oz) pineapple tidbits
1/4 C slivered blanched almonds
3/4 C 1/2 in. x 2 in. Oriental Gluten Strips (page 50)
Sausage seasoned baked gluten meat balls (page 18) can be
used in place of strips.
salt and pepper to taste
(continued)

Melt butter, add curry, celery, onion, green pepper and sauté until vegetables are tender. Blend starch with syrup drained from pineapple and add water to make 1 C. Add sugar, vinegar, soy sauce and Worcestershire sauce.

Add orange juice and pineapple. Cook to a boil for 1 min, stirring constantly. Stir in gluten and almonds. Simmer 2-3 min. Add salt and pepper.

Serve with hot rice.

Light, Fluffy Brown Rice

1 C brown rice
2 C chicken or vegetable stock or water

Place rice in casserole dish. Pour boiling liquid over rice and stir. Cover, place in preheated oven at 350 for 1 hr.

Mushroom Stroganoff *Using Gluten Strips*

1 1/2 C cooked and seasoned gluten strips
1 egg
1/4 C flour
2 T butter or oil

Sauce:
1 C mushrooms, sliced
1 onion, grated
2 T butter or oil
3 T flour
1/2 tsp basil
1/4 tsp salt
1/8 tsp white pepper
pinch of nutmeg
1/4 C water
1 C sour cream

Coat firm gluten strips in egg, then flour. Brown in skillet
with butter or oil. Remove from skillet. Sauté mushroom
slices and onion in butter or oil. Cook 3 or 4 min or until
onion is barely tender. Stir in flour, then add remaining sauce
ingredients except for sour cream and cook over low heat for
2 minutes, stirring constantly until the mixture thickens.
Cover and simmer an additional 2 minutes. Add sour cream.
Serve on cooked noodles, rice or baked potatoes.

Chicken Stroganoff

Sauté:
1/4 C green peppers, chopped
2 T onion, chopped
2 T butter or margarine

*Mix cream soup and sour cream together, then add to remain-
ing ingredients and add to the above mixture:*
1/2 C sour cream
1 C cooked, diced or strips of chicken flavored gluten.
1 can Cream of Mushroom soup
2 C egg noodles, cooked, wide or medium.
1/2 C sliced olives
1/2 tsp paprika sprinkled on top

Bake in a 1 quart casserole for 30 min. at 350°.

Jerky Strips

3 C unseasoned raw gluten,
Roll out on wetted surface 1/8" to 1/4" thick. cut into jerky-
size strips 1 to 1 1/2" wide and 3" long. (try scissors or pizza
cutter)

Mix broth ingredients and bring to boil. Drop gluten pieces
into broth (see next page), one at a time, and simmer 20 minutes, stirring occasionally.
(continued)

Broth
1 C soy sauce
1 1/2 C water
1 tsp Worcestershire sauce
1 tsp liquid smoke
1/3 C honey
1 tsp pepper
1/2 tsp onion powder or salt
1/2 tsp garlic powder or salt
1 tsp Basic Sausage Seasoning (page 100)
1/2 tsp Tabasco sauce (optional)
4 T Kitchen Bouquet - enough to darken - (optional)

Place pieces on cookie sheet that has been coated with a non-stick baking spray. Place in 300° oven (leave door ajar slightly) until pieces appear dry on top. Turn pieces over then let remain in oven until texture is very firm, about 30 min. or dry in a food dehydrator or in the hot sun for about 2 hours.

Makes approximately 24 strips.

Note: Jerky pieces will often vary slightly but texture becomes even when stored in plastic bags. Can keep in jar at room temperature if pieces are thoroughly dried, otherwise – freeze.

SEAFOOD-FLAVORED GLUTEN DISHES
Use any fish flavored soup base that is available. Check with restaurant supply stores or see page 96.

Clam Chowder

For "Clam Seasoned Gluten," place 1 C steamed, firm gluten cubes in 1 C water and 1 T clam base.

Cook in sauce pan until tender:
1 C water
1/2 C onions, chopped fine or grated
1 1/2 C potatoes, finely chopped
1 1/2 C celery, chopped fine

Add to:
4 C white sauce (page 97)
3 1/2 T clam base
Add:
Clam seasoned gluten pieces. Heat through and serve with wheat rolls.

Gluten "Crab" Pieces

Simmer 1/2 C steamed gluten, torn in small pieces to resemble crab in 1 C water and 1 T crab base until liquid is boiled away. (see page 14)

Cottage Cheese and Crab Salad

Combine the following:
1/3 C Gluten crab pieces, broken up small
1/2 C mayonnaise
4 T cottage cheese, dry curd
Chop finely and add to above mixture:
1 green onion
1 1/2 C celery
1/4 C green pepper
1 T sweet potato, raw, grated
Refrigerate and serve on lettuce leaf, in tomato (middle scooped out and added to salad,) as a sandwich filling or in pocket bread with slice of Swiss cheese and sprouts.

Oriental Crab Salad

Mix together:
1/2 C mayonnaise
1 tsp soy sauce
1/8 tsp ginger powder
1/2 tsp lemon juice
Add:
1/2 C Gluten Crab pieces, broken up smaller
1 C bean sprouts
(continued)

Chop very fine and add to above mixture:
1 green onion
1 C celery
2 T sweet red pepper
1 C celery
1 C water chestnuts or jicama

Refrigerate and cover until served. Could mix in chow mein noodles if desired. (**Note:** This could also be made with Chicken Gluten pieces.) Garnish with fresh parsley. Could arrange sliced boiled egg on top.

Crab Salad or Sandwich Filling

1 C Gluten "Crab" Pieces
1-2 boiled eggs (opt.)
mayonnaise, thinned with pickle juice, to moisten
1/4 C diced gherkins
1 chopped green onion
salt and pepper to taste

Mix all ingredients. Spread on bread. Include lettuce for a very tasty cold sandwich. (Note: For a Ham Salad Sandwich, use ham flavoring.)

PET FOOD

Pets thrive on a varied, wholesome diet just as we do. The main difference is that they need more protein and calcium. Their foods should include fresh meat or gluten, dairy products, fish, vegetables, nuts, eggs, grains and legumes. (They do not have to worry about cholesterol - but the preservatives, artificial colorings, and below standard meat by-products are bad for them.)

Spend time with your pet. They always want to please you. They will thrive on the love you give them and the nutritional recipes below.

Gluten Mix for Pets

Add 4 T powdered non-instant milk or soy flour to the recipe for Ground Gluten Pieces (page 17)

Mix ingredients. Spread on baking sheet and bake at 350° until firm, about 20 minutes. For a crunchy texture, omit oil and bake to desired hardness.

Extra Nutrition

For a prettier coat, healthier pet and less fleas, add 1-2 T of the following mixture to your pet's food every day.

1 1/2 C nutritional yeast
1 C wheat germ
3/4 C bran
3/4 lecithin granules or powder

As an extender

add 3 C Cooked Ground Gluten Pieces to 15 oz. can any pet food

Jerky for Pets

Use the jerky recipe (page 53), making the pieces more firm for them. Omit the hickory smoke. Can dry harder, if desired.

Dog-Gone Good Biscuits That are the Cat's Meow

(great for cats, dogs and masters)

Pet Biscuit #1

Slightly beat together:

1 large egg
1/3 C vegetable oil
1/2 C milk or gluten water
1/2 tsp salt
4 T Soup base (chicken, beef, fish) or any of the sausage seasonings (page 16, 100-101)

(continued)

Add:
1 1/2 C whole wheat or multi-grain flour
1 C dry skim milk
1/3 C quick-cooking oats
2 T wheat germ (optional)

On a lightly floured surface, knead ball one minute. Work on more flour if it sticks to your hands. For different shapes, use any of the following methods:

• Form into small 2 inch balls and press with fork onto a greased cookie sheet.
• Roll out to 1/2 inch thick and cut with cookie cutters and place on greased cookie sheet.
• Roll dough on to a greased cookie sheet to 1/2 thick and score into squares etc. with pizza cutter.

Pet biscuit #2

Preheat oven to 350°. Cream together the following.
6 T butter or Vegetable oil
1 egg

Mix together:
2 1/2 C whole wheat flour
1/2 C non instant powder milk
1/2 tsp salt
2 tsp Nutritional yeast
2 tsp wheat bran

Add flour mixture and 1 C water alternately to egg and butter mixture. Knead 2-3 minutes and roll out to 1/4 to 1/2 inch thick on floured surface. Dip cookie cutter in flour, cut and bake 25-30 min.
Leave out in air for hard biscuit, store in airtight container for soft ones.

USING THE STARCH AND GLUTEN WATER

Recipes for the Saved Water, Starch and Bran (page 12)

Pizza Crust

2 C starch and gluten water
2 C flour
4 tsp cream of tartar
1 tsp soda
1 tsp salt
5 T oil

Mix all ingredients (oil last). Spread dough with hands or rolling pin on pizza or baking sheet. Makes 4 12-in. crusts. Bake at 350° for 10 min or until dough is set but not browned, before garnishing with sauce, cheese, gluten sausage, etc. Bake again until cheese melts and crust is heated through.

Note: Crust will be too tough if dough is thick. Be sure to keep it workable by adding a little water if necessary. This recipe also works well for a tender cracker.

Crackers

1 C starch and gluten water,
1/2 C flour
1 tsp cream of tartar
1/4 tsp soda
2 T oil

Mix all ingredients (oil last). Cracker batter should be consistency of thin cake batter. (If crackers come out too tough or thick, batter was not thin enough. Add more water if necessary.)

Spread on oiled 12 x 14 inch cookie sheet. Get as even as possible—about 1/4-inch or less, tip pan side to side. . Sprinkle salt on top of batter.

Bake at 400° for 10 min. Take out of oven when firm, Score in squares with fork handle; mark holes with tines. Return to bake 10 min; turn over and bake additional 5 min or until crisp. Break and store.

For variations, add garlic or onion salts, herbs, spices; top with powdered cheese or sesame seeds.

Thickenings

Pour off all the water from the settled starch. (This procedure may be repeated 3 times, letting the starch settle each time for a few minutes between pourings, or overnight.)

Gravy

4-5 T thick starch to 2 cups liquid

Stew

6-7 T thick starch to 2 cups liquid

Other

Thicken according to recipes.

Note: Thickening with wheat starch gives about the same results as thickening with corn starch. It is more clear than flour and works well for sweet types of thickenings.

ICE CREAM USING THE SETTLED STARCH (page 12)

Vanilla

1/2 C raw sugar
1/4 C honey
3 T starch
4 egg yolks
2 tsp vanilla
Dash of salt
1 1/2 C yogurt, whipped

Mix sugar, honey, starch water. Boil 3 min stirring constantly.
Blend with egg yolks, salt and vanilla in blender. Mix into
yogurt. Pour into a refrigerator tray. Cover with waxed paper.
Freeze 2-3 hours.

Raspberry

1 T lemon juice
1 T protein powder (health food store)
2 T starch
1 10 oz pkg frozen raspberries
2/3 C sweetened condensed milk
1 C yogurt

Put all ingredients into blender except milk and yogurt. Mix
20 sec. Slowly add milk and yogurt. Freeze.

Ice Cream Tiger

1 T Tiger's milk (protein powder)
1 C starch
2 C frozen fruit
1 tsp vanilla
1 rennet tablet (Junket) at grocery store

See rennet "Junket" instructions for freezing.

Mocha Carob Ice Cream

1 C reconstituted powder milk
1 T parched wheat (page 205)
2/3 C evaporated milk
1 T instant chocolate drink mix
1/2 C starch
1 T carob powder or instant chocolate drink mix
4 T melted dipping carob
2 T raw sugar
1 T honey
1/4 tsp maple flavoring

Simmer parched wheat in reconstituted milk 10 min or until evaporated down to 1/2 C. Strain out wheat and freeze milk Put in blender all other ingredients and mix at high speed for 10 seconds. Add frozen parched wheat and milk mixture. Mix 30 seconds or until smooth. Pour into container and freeze. Can be whipped again before serving.

Variation: Omit parched wheat and add mint flavor, postum or malted milk to taste.
Hint: To add a special touch, top with Gluten Crunch.

FRUIT DRINKS
using gluten water (page 12)

Basic Recipe

2 C gluten water
2 C frozen orange juice (reconstituted with gluten water)
1-2 C fresh or frozen fruit (when using fresh fruit, add ice to create desired thickness)

Add ingredients to blender and mix 1-2 minutes or until smooth.

Basic Recipe for Bran Flakes:

2 C raw bran (page 13)
1/2 tsp salt
2 T sugar (optional)
water to thin batter

Mix ingredients. Pour paper thin on oiled cookie sheet. Tip pan from side to side until batter is fairly even. Bake 20 to 25 min at 300°. Lift sheet of bran from pan immediately. Let cool. Break into pieces.

Variations:
• 2 C raw bran (page 13)
2 T protein powder or 1 T malt powder
1/2 tsp salt
2 T raw sugar
Mix and bake as in basic recipe

Cinnamon-sugar

Sprinkle cinnamon-sugar on top of batter after it has been in the oven 10 min.

Unsweetened coconut

Sprinkle coconut on top of batter after it has been in the oven 10 min.

PREPARING AND STORING GLUTEN MEALS AHEAD

Raw Gluten

By making 5 batches of gluten using 12 C flour for each, the amount of raw gluten derived will be about 14 C. Follow various instructions and be prepared to freeze or cook the gluten within 24 hrs, as it will not keep at this stage. The next steps will give suggestions on how to prepare the 14 C of raw gluten into some basic preparations.

Ground Gluten

Steam 8 C raw gluten, Grind in a meat grinder or food processor and you have ground gluten. Store in plastic bag in fridge. Yield: 18 C ground gluten.

Prepared Gluten

Prepare Ground Gluten in any of the following ways:
Meat balls   Sausages
Patties   Ground Gluten Mix
Candies   Cereals
Store in baggies or airtight containers in freezer.

Gluten Steaks

Steam 4 C of the raw gluten to make 6-7 steaks. Can freeze for future use.

Basic Yields:
12 C whole wheat flour makes from 2 1/2 to 4 C raw gluten depending on protein in flour.
4 C raw gluten cooks into 9 C ground gluten
9 C ground gluten (cooked) = 3 Lbs hamburger (cooked)

12 C whole wheat flour made into gluten produces:
- 512 gluten cubes - (1/2 x 1/2 in.)
- 150 meat balls
- 20 steak slices - (4 x 1/2 in.)
- 4 pie crusts (like graham cracker crusts)

QUESTIONS MOST OFTEN ASKED

Refer to "The Quick Wholesome Foods Video" to solve many of the questions asked. Seeing the textures and techniques makes gluten making easy.

Q. What Kind Of Flour Do I Use To Make Gluten? Can I Use The Kind Sold At The Markets? What Texture Should It Be?
> A. Whole wheat flour is best because the valuable bran can be reserved. White flour is usable; however, the gluten from it is harder to control in the raw stage. The texture of either flour should be fine to medium fine. Yes, commercial whole wheat flour can be used, but it is usually made from a softer wheat containing less protein. The yield may be less.

Q. Is It Necessary To Cook The Gluten Before Putting It Through The Food Chopper To Make Ground Gluten?
> A. Raw gluten will not go through a food chopper. It must be cooked (steamed) first.

Q. Is Gluten A Complete Protein?
> A. Gluten is low in the amino acid lysine. A combination of gluten and foods high in lysine is a very beneficial protein.

Q. Is Gluten Fattening?
> A. Any protein has a certain amount of fat, but vegetable proteins usually contain about one-fifth as much fat as beef.

Q. Should I Use Hot, Warm Or Cold Water To Rinse The Gluten?
> A. Water should be cool or comfortable to hands.

Q. What Sort Of Texture Makes Good Gluten Steaks?
> A. The best texture resembles bologna, a poor texture is like a slice of bread.

Q. How Do I Test The Mixed And Rested Dough To See If The Gluten Is Developed And Ready To Be Rinsed?
> A. Take a small handful from the dough, hold it under running water, squeeze and work it. The gluten is ready when it has elasticity and holds together. If it does not the texture was too thin or too thick (see "The Quick Wholesome Foods Video"), or the wheat is soft and needs a longer rest time (30 minutes to an hour or longer).

Q. I End Up With So Much Less Than When I Started After Rinsing The Gluten. What Am I Doing Wrong?

A. Most likely nothing. Expect to have less gluten than the dough you started with, especially if the wheat flour is soft.

Q. How Much Water Does It Take To Rinse The Gluten? It Seems Like A Lot!

A. About one gallon or more for 12 cups flour.

Q. What Is The Reason For Rinsing The Dough?

A. Rinsing brings just the protein cells together. Unless the other components are taken away in rinsing, the elastic quality cannot be developed.

Q. When I Freeze The Gluten That Has Been Steamed In A Container, Do I Leave It In The Container Or Take It Out?

A. Take the gluten out and slice it or prepare in a variety of ways. Place in baggy or large freezer container and then freeze.

Q. How Long Can I Store The Flour And Water Mixture Before Washing It Out?

A. Not more than 24 hours, as it ferments (resembling raised yeast) and it loses the ability to hold together during the rinsing process.

Q. What Can I Do With Cooked Gluten That Was Spongy Rather Than Firm And More Solid.

A. Steam longer or grind it to use it in ground gluten recipes.

Q. How Long Will Gluten Store?

A. The reserved starch and water - 5-7 days in fridge in an air-tight container. If gluten starts to ferment, it falls apart. Cooked gluten, nothing added but seasonings — 1 week to 10 days.
Gluten residue, starch water — 1-2 days. Bran — 3-5 days.

Q. How Do I Use Gluten As A Meat Extender?

A. Use only one-fourth ground meat, the rest Ground Gluten. Omit the beef or other meat seasoning bases. Prepare as your recipe indicates.

TERMS USED IN THE BOOK

Baked Sausage, Chicken or Beef Mix
> Baked or steamed gluten which has been ground, other ingredients added, including seasonings and now ready to cook in various methods.

Crushed or Minced Clove of Garlic
> Several types of inexpensive garlic presses are available in most hardware departments, or the garlic can be pulverized with a hammer, or minced finely with a paring knife.

Dipping or Chunk Carob
> The same as dipping chocolate, but made from the carob bean, a healthy alternative which does NOT contain caffeine. Can be found at health stores, candy shops and many grocery stores.

Firm Gluten
> Steamed to bologna-type texture. Hint: Before cooking the raw gluten using any method, let rest a few hours in the fridge.

Gluten Balls
> Same as a meat ball, using ground gluten in place of the hamburger

Gluten Cubes
> Cooked gluten cut into cubes. Usually prepared by boiling or steaming method.

Gluten Steaks or Steaklets
> Raw gluten that has been steamed and sliced

Gluten Strips
> Raw gluten which has been rolled out thin, cut into strips and boiled in a broth then dried out in the oven or dehydrator for a chewy texture. They can be used for marking jerky, stir fry or stroganoff.

Gluten Water, reserved

The liquid which has been reserved from the rinsing process in isolating the gluten. Gluten Water contains the water-soluble vitamins from the whole wheat especially the B vitamins. Use by reserving and then adding to other foods.

Ground Gluten

Raw gluten which has been steamed and put through a meat grinder or food chopper to give a ground hamburger-like texture.

Ground Gluten Pieces

Ground gluten which has been mixed with other ingredients, such as: egg, flour, oil and seasoning, then baked in oven until firm.

Powdered Milk

Dehydrated nonfat, non-instant or instant dry milk. (Instant is sweeter.)

Prepared Gluten

Cooked seasoned gluten (using any method) and ready to use in recipes or meals.

Raw Gluten

The bubblegum-like texture which is left after rinsing the rested flour and water mixture.

Reserved Starch and Bran

Products from the Gluten-making process that have been separated and saved.

Rinsing the Gluten

Working the mixed flour and water, which has rested, with clear water until the extra properties are rinsed away from the bubble gum-like gluten which holds together.

Softened Dehydrated Fruits

Dried fruit placed in a jar, water poured in and immediately out again, then stored in airtight container and refrigerated. The fruit should be chewy and soft. Leave out in the open if it is too moist.

Stirred Gluten (Quick Method)

The flour and water in a large bowl are stirred together with a large spoon. After letting mixture rest 15-30 minutes, mixture is ready for the rinsing process.

Washing the Gluten

The same as rinsing.

Gluten Flour

A commercially prepared gluten flour is available at health food stores and bulk outlets. All you need to do is add water. However you lose the benefit of the starch, minerals, vitamins and bran.

You will find some delicious and interesting new taste treats as well as familiar and favorite seasonings. If you are interested in eliminating monosodium glutamate, artificial colorings and flavors, unnecessary sweeteners, meat fats and preservatives, you will especially appreciate this section. Adjust these recipes to suit your family's tastes.

SECTION 2

SEASONINGS AND MIXES

Seasonings and Flavorings

The remarkable properties of herbs and spices have been appreciated by nearly every civilization since ancient times. They regarded herbs as being endowed with numerous almost-magical properties which made them valuable in cooking, medicines, cosmetics, perfumes, dyes and embalming. The popularity of both herbs and spices is on the rise again today for several reasons. More and More people are enjoying good cooking and since flavor is a vital component of good food, herbs and spices are called upon increasingly by the creative cook.

The use of natural herbs and spices is suggested in making your own seasoning. These simple and quick methods can save you time and money.

Information for preparing Herbs for recipes

*Many of the herbs called for in the seasoning blends come only in seed or leaf form. For this reason you will need some method of powdering them. The best way is with a small **nut mill** (also called coffee grinder). Another way is a strong blender; however, it does not powder them as well.*

When buying in bulk or for hard to find ingredients, check your local co-op outlets, restaurant supply store or wholesale bakery and kitchen stores.

HERB AND SEASONING GUIDELINES

Don't be afraid to experiment. While some herbs and spices seem traditionally to go with certain foods, there's no "right" herb or spice for any food. It's right if you like it. When trying a new herb, crush some in the palm of your hand, let it warm there for a few moments, then sniff and taste it. You can get it a fairly good idea of what it'll taste like when used in cooking.

General Rules

Following are a few general hints for using herbs and spices:

1. Start with a light touch until you discover the amount that suits your family tastes. Measure and/or taste as you go.

2. Don't mix too many different herbs or those with conflicting or strong flavors in one dish (dillweed with caraway, for instance): they'll fight and cancel each other or produce strange off-flavors.

3. Limit the number of herb dishes in a meal-perhaps to one or two.

4. Adding seasonings to food:

Ground spice - add at the same time as salt. In batters and dough, sift with the flour for even distribution.

Whole spice - add at the start of the cooking period so that the heat can draw out the flavor.

Long-cooking foods - add during the last hour of cooking so that the flavor will not be lost by prolonged heating.

Cold dishes and beverages - add during preparation and let stand awhile.

Cold dishes with whole spice - heat whole spice in juice of food to draw out the flavor, then let stand in the refrigerator overnight.

5. Amount to Use: Generally, one part dry herbs or spices equals three parts fresh: about 1/8 teaspoon garlic powder equals 1 medium clove garlic. If you have no recipe, allow 1/4 teaspoon dry herb or spice per pound of meat or pint of liquid (except garlic and onion powders and red pepper, which should be less-about 1/8 teaspoon).

When using the herb in leaf or powder form, THE LEAF IS DOUBLE THE AMOUNT OF THE POWDER. Example: If the recipe calls for 1 tsp rosemary leaves, use 1/2 tsp powdered rosemary.

Frozen Herbs
Requires about twice the amount you would use fresh.

Stored in oil or vinegar
Can be used in same amounts

Fresh Herbs
Triple the amount of fresh herbs used in place of dry herbs or if a wilder quality is desired (i.e.. rosemary, garlic, etc.). Use 1 T fresh to every 1 tsp dried when substituting.

1 small onion = 1 tsp dried minced onion
1/4 C chopped fresh onion = 1 T dried minced onion
1 clove garlic = 1/8 tsp garlic powder
1 C tomato juice = 1/2 C tomato sauce plus 1/2 C water.
1 tsp dry mustard = 1 T prepared mustard

Basic Rule
Add some, taste and add more if desired.

6. Buying The most often asked question is **Where do I find these herbs?** Check grocery stores, restaurant supply, wholesale or discount outlets, co-ops, health food stores or Seventh-day Adventist stores or groups (great source for vegetarian seasonings). Buy spices and herbs in *small* sizes or amounts so you can use them up while they still have top favor and aroma. They should look fresh and have bright color. And buy from a market where there's a rapid turnover. Long storage (whether at home or at the store) deteriorates most herbs and spices. Herbs are the most fragile. If possible, buy them in leaf form and crush just before using. Ground spices (cinnamon, pepper, cloves, ginger, etc.) come next. Whole spices and seeds will keep the longest. As you buy herbs and spices, date the jars or cans with grease pencil so you can keep track of freshness.

When buying in bulk (see resources in back of book) go in with others and divide the product. This way you not only save money but will have herbs that are more fresh.

7. Store herbs and spices in tightly covered containers away from heat. They keep best in cool, but not cold, places. As soon as herbs and spices lose their delicate, distinctive aroma they should be replaced because they have also lost their flavor.

SEASONING MAKES THE DIFFERENCE

Good seasoning is difficult to describe but wonderful to taste. The right seasonings added to food make it an adventure in good eating.

The ability to season food properly begins with the choice of the right products. Before you select yours, here are some things you'll want to know.

The term spice is often used to refer to over thirty different spices, herbs, and seeds. Each has its own personality and gives a subtle deliciousness to even the simplest and least expensive foods.

Generally speaking, use spices with a light hand. Their function is to improve the natural flavor of the food, not to mask it or cover it up.

Spices come from roots, bark, stems, leaves, buds, seeds, or fruit of aromatic plants which usually grow in the tropics. Some are sweet, some "spicy-sweet", some are "hot." The majority of them are used in the ground form.

Herbs are the leaves of plants which grow only in the temperate zone.

Fresh Herbs: Keep fresh herbs flavorful by soaking them in olive oil; then store in the refrigerator.

Frozen Herbs: An ideal way to have an all-year-round herb garden. Strip all heavy stems from herbs; wash carefully; drain, then let dry on towels at room temperature 1-2 hours or till free of moisture. Freeze herbs chopped or whole. Place chopped herbs in containers with easily removable covers so you can readily spoon out what you need. Whole herbs can be frozen in plastic bags.

Fines Herbs: a combination of parsley, chives and tarragon.

Seeds used in cooking are the seeds and also the small fruit of plants which grow in both tropical and temperate zones.

Blends are mixtures of several spices combined for use in certain types of cooking. Examples are: poultry seasoning and pumpkin pie spice.

Seasoning salts such as garlic salt and celery salt are mixtures of salt with ground seasoning.

Parsley lends its freshness to other herbs. Try mincing equal amounts of fresh parsley with such herbs as dried dill, basil, marjoram, or rosemary. The taste is great—so fresh!!

Note: Nutritional Yeast - a powdered or flaked Brewers yeast listed in my seasoning mixes. This gives the blends of spices a mellow and rich flavor, along with the added nutritive value

A BASIC GUIDE FOR SEASONING WITH HERBS

Anise — The seeds are most commonly used in teas, and baking cookies and cakes. The leaves are often used in salads.

Sweet Basil — Basil is widely used in all types of dishes, such as shrimp, broiled fish, stew, hash, meat loaf, tomatoes, eggs, beans, cheese, and in appetizers.

Caraway — Caraway seeds add zest and flavor to vegetables when baked or cooked, such as cabbage, sauerkraut, carrots, onions, baked potatoes, soups and many other dishes.

Chives — Chopped in salads, omelets, sour cream, soups and cheese dishes, it gives added zest and color.

Dill — Fresh dill leaves are especially tasty in cucumber salads, cottage cheese, coleslaw, vegetable salads, and egg salads. Also, with all meats, potatoes, peas, beans, tomatoes, and spinach. The whole plant, excluding the roots is used in making pickles, and seeds are also used for vinegar, seed cakes and bread.

Fennel — The stems of this herb can be used like celery. Seeds are used in teas and bread. The seeds and leaves add a licorice flavor to soups, chowder, fish, pickles, and fish sauces.

Garlic — The most common uses of garlic are in cooking, in salads, soups, garlic butter, sauces, and Italian dishes.

Lemon Balm — The dried leaves are widely used for tea, in salads, and in all meat dishes.

Sweet Marjoram — Marjoram leaves are used in salad dressings, salads, vinegars, soups, cooked vegetables, and in meats.

Mint — Mint is very popular as a tea, and the leaves can be used fresh or dry. It may be used for mint sauces, in salads, drinks, sprinkled on fresh fruits, and mixed with other herbs such as Lemon Balm or Alfalfa for a delicious, refreshing tea.

Parsley — Parsley has more value than just as a garnish. It is very rich in Vitamins A and C. It is also rich in Iron. Parsley should be used in all salads, soups, etc.

Rosemary — Its uses are many and varied, from using sprigs as cut flowers, the fresh leaves for flavoring soups, stews, vegetables, meats, and tea.

Sage — The leaves are dried for later use for meats, fish, and with some vegetable recipes, dressing (stuffing.)

Savory — The leaves of savory are especially good with green beans, and yellow beans. Also it can be used in salads, egg dishes, sauerkraut, cabbage, ground beef dishes, and fish.

Shallots — Shallots look like small onions and are used in the same manner, but in smaller quantities. It may be used in fresh peas, salads, salad dressings, meat loaf, red beet tops cooked, and with many other types of cooked greens, and in baked dishes.

Tarragon — Fresh leaves are used for Tarragon Vinegar, tartar sauces, salad dressings, in salads, poultry, cheese dishes, egg dishes, and Orange Bread.

Thyme - It is used in stuffings, salads, salad dressings, soups, tomato and cheese dishes, and flavoring of all types.

WILD HERBS FOR SEASONING
for the outdoor enthusiast

Chamomile (leaves and flowers)   in cooked cabbage and broccoli.

Carrot, Wild (seeds, leaves)   a salt substitute, in soups, with fish.

Catnip (leaves, flowers)   a substitute for peppermint, in sauces.

Clover (leaves, flowers, fruits)   soup and stews.

Garlic, Wild (whole plant)   soup, salad dressing, sauce.

Juniper (fruits)   in vinegars, soups, stew with venison.

Onion, Wild (whole plant, seeds)   soup, dressing, sauce.

Peppergrass (leaves, fruit)   herb salt, soup, stew, salad, vinegars.

Peppermint (whole plant)   jelly, soup, lamb, potatoes.

Shepherd's Purse (whole plant)   herb salt, soup, stew, salad.

Spearmint (whole plant)   same as peppermint.

Tansy (young leaves)   a substitute for sage.

Watercress (leaves)   salad, drinks.

Wormwood (leaves, flowers)   in vinegars, a discreet substitute for tarragon.

SEASONING FOODS WITH HERBS AND SPICES

APPETIZERS

Hors D'oeuvres   chervil, oregano, paprika, parsley

Cheese dips   basil, dill weed, oregano, thyme,
and spreads   chervil, marjoram, sage, parsley, summer
                          savory, tarragon

Deviled   tarragon, dill weed, marjoram, curry powder, or
stuffed eggs   summer savory

Dips   curry powder, oregano, chervil, parsley

Mushrooms   oregano, marjoram

Seafood cocktails   basil, dill weed, thyme, bay leaves, tarragon
and spreads

VEGETABLES

Asparagus   lemon peel, thyme

Broccoli   lemon juice, onion

Brussels sprouts   lemon juice, mustard

Cabbage   dill weed, caraway seeds, oregano, lemon juice,
                          vinegar, onion, mustard, marjoram

Carrots   marjoram, ginger, mint, mace, parsley, nutmeg,
                          sage, unsalted butter, lemon peel, orange peel,
                          thyme, cinnamon

Cauliflower   rosemary, nutmeg, tarragon, mace

Celery   dill weed, tarragon

Cucumbers   rosemary, onion

Green beans	tarragon, basil, dill weed, thyme, curry powder, lemon juice, vinegar
Peas	mint, onion, parsley, basil, chervil, marjoram, sage, rosemary
Potato	dill weed, basil, savory, tarragon, bay leaves, chervil, mint, parsley, rosemary, paprika, mace, nutmeg, unsalted butter, chives
Spinach	basil, oregano, chervil, marjoram, mint rosemary, mace, nutmeg, lemon, tarragon
Squash	basil, mint, chervil, oregano, thyme, sage, saffron, ginger, mace, nutmeg, orange peel
Tomato	basil, tarragon, oregano, dill weed, bay leaves, chervil, curry powder, parsley, sage, cloves
Zucchini	oregano, basil, savory, rosemary, marjoram, mint saffron, thyme

ENTREES

Eggs and cheese	ground sage, marjoram, tarragon, thyme, chervil, curry powder, mace, parsley flakes, turmeric, oregano, rosemary, sage, saffron
Beef	basil, bay leaves, cumin, oregano, dill weed, tarragon, thyme, sages, savory, rosemary, garlic, mustard, mace, ginger, curry powder, allspice, lemon juice, pepper
Fish and shellfish	bay leaves, tarragon, thyme, dill weed, basil, chervil, marjoram, oregano, parsley, rosemary, sage, lemon peel, celery seeds, cumin, saffron, savory, dry mustard
Poultry	bay leaves, rosemary, oregano, savory, basil, saffron, sage, dill weed, marjoram, tarragon, thyme, paprika, curry powder, orange peel, cranberries, mushrooms

Pork   basil, thyme, oregano, dill weed, marjoram,
                        cloves, garlic, ginger, mustard, nutmeg, paprika,
                        rosemary, savory, curry powder, apples

HERBS FOR EGGS AND CHEESE

Cheese Casseroles: Any cheese dish will respond to a dash of sage, marjoram, basil, dill or thyme.

Cheese Rarebit: Include basil and marjoram in the cheese sauce, or some thyme.

Cheese Sauce: Add 1/2 tsp mustard for zest or try a very sparing amount of tarragon. In a sauce for vegetables, marjoram is just the thing.

Cheese Soufflé: Add basil to suit your taste.

Cottage Cheese: Add onion, salt, dill or caraway seed.

Cream Cheese: Blend in basil or parsley flakes. Spread on thin rye or use as dip.

Cheese Spread: Season snappy soft cheese with thyme and celery salt.

Deviled Eggs: Add savory and mustard.

French Omelet: Dash of basil in the eggs.

Scrambled Eggs: Shirred Eggs: Sprinkle lightly with savory or use small amount of rosemary.

Soufflé: Add from 1/4-1/2 tsp marjoram to 4 eggs. Serve with hot tomato sauce.

Tomato (Spanish) Omelet: Just a bit of oregano, basil, bay leaf, saffron, thyme, sage, or savory.

FRUITS AND DESSERTS

Apples	allspice, cardamom, ginger, cinnamon, cloves, nutmeg
Bananas	allspice, ginger, cinnamon, nutmeg
Oranges	allspice, cinnamon, anise, nutmeg, cloves, ginger, mace, rosemary
Pears	allspice, cinnamon, nutmeg, anise, mint
Fruit compotes	basil, rosemary, saffron, thyme
Puddings	arrowroot, cinnamon, cloves, lemon peel, vanilla bean, ginger, mace, nutmeg, orange peel

BEVERAGES

Cranberry cocktail	cinnamon, mint
Fruit juices	marjoram, mint, saffron, cloves, lemon and orange peel
Iced tea	cinnamon
Milk drinks	cinnamon, nutmeg, vanilla bean
Spiced cider	ginger
Tea	sage, mint
Tomato juice cocktail	allspice, basil, tarragon, cloves, oregano, lemon peel, summer savory

HERB SEASONINGS FOR SALADS

THESE HERBS	*ADD ZEST TO*
Basil	Tomato and green salads; fresh tomato slices
Caraway	Coleslaw; beet and potato salads
Chives	Potato, cucumber, mixed vegetable and green salads
Dillseed	Coleslaw; potato and cucumber salads
Marjoram	French dressing; mixed green and chicken salads
Mint	Fruit, cabbage and celery salads; coleslaw
Mustard	Potato salad; French and oil/vinegar dressings
Oregano	Potato, mixed green and seafood salads; tomato aspic
Parsley	Greens, vegetables, shellfish salads; garnish for all salads
Rosemary	French dressing or mayonnaise for chicken or potato salads
Savory	Mixed greens, potato and tomato salads; green vegetables, aspic
Tarragon	Allspice; tomato and beet salads; mayonnaise and herb dressings

HERBS AND SPICES *to flavor the following*

Italian: Oregano, rosemary, marjoram, basil, thyme, sage, garlic,

Mandarin Sweet & Sour: cornstarch, salt, garlic, sesame seed, ginger, chili pepper, Worcestershire powder, fennel, red pepper, paprika, cinnamon

Mexican dishes: Onion, chili powder, cumin, garlic, oregano, coriander

Oriental: Sweet Blend - cloves, anise, fennel, licorice root, cinnamon (black pepper optional).

Soups and Vegetables: Onion, celery seed, thyme, basil, bay leaves, pepper, garlic, parsley

Stir-Fry: ginger, soy sauce, Oriental sesame oil, beef, ham or chicken base, sesame seeds

Sweet Blend: grated orange or lemon peel, cinnamon, nutmeg, cloves, ginger

Teriyaki: Powdered miso, salt, flour, onion, mustard, garlic, vinegar, honey or sugar

Other items needed in making sauces, flavorings etc: Tamari (soy) sauce, Liquid or pwd natural Hickory Smoke, Worcestershire Sauce, Kitchen Bouquet (browning and seasoning sauce), Marinades, Flavored extracts such as vanilla, mapleine, lemon, orange, cornstarch etc.

SWEET SPICES

Replace Sugar with Spices for Delicious Desserts
In many dessert recipes, it is possible to cut the amount of sugar by as much as one-half. Certain spices called "sweet spices" can take up the slack in flavor when sugar, and calories are cut down. This group of seasonings is not markedly sweet by themselves, but their flavors are especially compatible with sweetness.

Sweet Spice

In a bowl, combine 1 T each of cinnamon and ground ginger, 1 1/2 tsp each of allspice and nutmeg, and 1 tsp ground cloves. Store in an airtight container and use it to flavor gingerbread and other desserts.

All Spice Substitute

For 2 tsp allspice.
1 tsp cinnamon,
3/4 tsp nutmeg
1/4 tsp cloves

Ginger Sugar

Into a bowl, sift together 2 C sugar and 1/4 C ground ginger and combine the mixture well. Transfer the mixture to an airtight container and let it stand for 2 or 3 days. Use the sugar to flavor cakes, puddings, squash, and applesauce and for glazing hams.

Vanilla Sugar

Bury 2-4 vanilla beans, halved and split lengthwise, in a large jar of confectioners or granulated sugar and let the mixture stand for 3-4 days. Used vanilla beans, rinsed and dried may be used. Sift the sugar over cakes and cookies and use it to flavor dessert batters.

Mexican Cinnamon Sugar

Into a bowl, sift together 1 C fine granulated sugar and 1 tsp each of cinnamon and cocoa. Store the sugar in an airtight container and sprinkle over warm puddings and custards.

Pumpkin Spice

4 T cinnamon
1 T powdered ginger
1/2 tsp powdered cloves
1/2 tsp nutmeg
1/2 tsp allspice
1 T cornstarch

Mix all ingredients and store till needed for pumpkin pie, squash dishes, sweet potato pie, gingerbread and spice cookies. Use 1 T Pumpkin Spice mix for each 9 inch pie.
Sift all ingredients together 3 times. Keep tightly stored at room temp. Keeps up to 3 months. Refrigerated it will keep up to a year. Makes 1/3 cup.

Homemade Sweet Spice Blend

Add 2 to 3 tsp of this mix to cake or quick bread batters; mix with ice cream, yogurt or fruit.
1 T finely grated orange or lemon peel
2 T ground cinnamon
1 T each ground nutmeg, cloves and ginger

Spread citrus peel on waxed paper. Let stand uncovered for 10 minutes. Mix peel thoroughly with spices. Store in an airtight container.

Apple Pie Spice

1 T nutmeg
3 T cinnamon
1/2 tsp powdered cloves
1/2 tsp allspice
1 tsp cornstarch

Mix ingredients together well. Keep tightly covered at room temp. for up to 3 months. Refrigerated for up to a year. Makes 1/4 cup.

SEASON SALTS

Basic Seasoning Salt

To reduce clumping put a cracker or few grains of rice in the shaker, keep from high humidity as much as possible and keep container closed when not in use.

2 C (1 lb) salt
1 T onion salt,
1 T garlic salt
2 T celery salt
2 T paprika
2 T dill seed
4 T black pepper
4 T white pepper
4 T white sugar

Thyme Salt

Up to 1 month ahead, with a mortar and pestle, crush 3 T thyme leaves to a fine powder or rub between fingers. In small bowl with fork, mix together thyme and 1/2 C salt. Store in tightly covered container. Makes 1/2 cup.

Oregano Salt

Prepare as above but substitute 2-3 T oregano leaves for thyme leaves.

Tarragon Salt

Prepare as above but substitute 2 T tarragon for thyme leaves.

Herbed Seasoning Salt #1

In a bowl, combine 1 C salt and 1 tsp each of dried parsley, chives, onion flakes, and summer savory. Store the salt in an airtight container and use it for roll or bread dough and to season salads.

Herbed seasoning Salt #2

Mix together in a jar;
1 tsp each: garlic powder, onion powder, and pepper
1 T each: thyme leaves and salt
2 T each: marjoram, oregano, and rosemary leaves
2 T each: basil and parsley flakes
3 T sesame seed.
Secure lid and shake until well blended. Makes 1 cup.

*Use this seasoning salt to make a quick herb-garlic bread or sprinkle it lightly on chicken, lean white fish, omelets, or vegetables before cooking.

Celery Salt

In a blender; combine 3/4 C coarse salt and 1/4 C celery seed, both toasted lightly in a dry skillet. Pulverize the mixture and sieve it into a small bowl. Store the salt in an airtight container and use it to season soups, chowders, and salads.

Bon Appetit Seasoning Mix

Excellent for dips, chicken salad, tuna spread, cole slaw, in sauces with butter and lemon juice for carrots.

Mix together:
4 T powdered fruit sweetener or 2 T sugar
3 T salt
2 T celery seed
3 tsp onion salt

Store in shaker container. Makes 3/4 C. Add a few grains of rice to keep from caking

LOW SODIUM SEASONING

Low Sodium Seasoning Salt

3 oz jar salt substitute
1 tsp celery seed
1/4 tsp onion powder
1/4 tsp garlic powder
1/2 tsp oregano
1/4 tsp thyme
1/4 tsp bay leaf, crushed
1/4 tsp black pepper
1/8 tsp ground anise seed

Combine all and store in tight container, shake the same as any salt seasoning.

Basil or Thyme

Add   1/4-1/2 tsp to 2 C green vegetables;
3/4-1 1/2 tsp to 1 1/2 lbs pork chops or roast;
1/8-1/4 tsp to 2 T butter or margarine for basting 1 lb fish or 1 1/2 lb chicken.

Chili Powder

Add   1-2 T to ground gluten, noodle or rice skillet dishes (about 8 C);
1/2-3/4 tsp to 8 C popped corn (1/3 C corn unpopped).

Curry Powder

Add   1 T to 2 lbs ground gluten;
1 1/2 tsp to 1 C uncooked long-grained rice;
1/2 tsp to tuna salad using 6 1/2-7 oz can tuna.

Dill Weed

Add   1/4- 1/2 tsp to 2 C green vegetables (even 3/4 tsp);
1/2-1 tsp to 4 C cooked noodles.

Dill Seed

Add   1/4-1/2 tsp crushed to 2 T butter or margarine for seasoning fish,
vegetables or bread.

Nutmeg

Add   dash -1/4 tsp to 2 C mixed vegetables, carrots, spinach;
1/8 tsp to 1 lb ground gluten;
dash - 1/8 tsp to 4 C creamed chicken or tuna.

Oregano

Add   1/4-3/4 tsp to 4 eggs for egg salad;
1/8-1/4 tsp to 1/4 cup butter for basting fish;
1/4-1/2 tsp to 2 C spinach, green beans or 3 C tomatoes.

Paprika

Add   1/2 tsp to 1/4 cup flour for dredging gluten;
1/2 tsp to 1/4 cup butter for seasoning white vegetables.

Parsley Flakes

Add   2-4 tsp to 4 Cups cooked noodles or 3 Cups cooked rice;
2 T to 2 lbs ground gluten;
1/4-1/2 tsp to 1/4 C butter for vegetables.

Tarragon

Add   1/4 tsp to 1 lb fish;
1 tsp to 3 lbs chicken;
1/4-1/2 tsp to 1/4 C butter for gluten.

INSTEAD OF SALT,
TRY SEASONING WITH HERBS AND SPICES

asparagus	lemon peel, thyme
broccoli	lemon juice, onion
brussel sprouts	lemon juice, mustard
cabbage	dill weed, caraway seeds, oregano, lemon juice, vinegar, onion, mustard, marjoram
carrots	marjoram, ginger, mint, mace, parsley, nutmeg, sage, unsalted butter, lemon peel, orange peel, thyme, cinnamon
cauliflower	rosemary, nutmeg, tarragon, mace
celery	dill weed, tarragon
cucumbers	rosemary, onion
green beans	basil, dill weed, thyme, curry powder, lemon juice, vinegar
peas	mint, onion, parsley, basil, chervil, marjoram, sage, rosemary
potatoes	bay leaves, chervil, dill weed, mint, parsley, rosemary, paprika, tarragon, mace, nutmeg, unsalted butter, chives
spinach	chervil, marjoram, mint, rosemary, mace, nutmeg, lemon, tarragon
squash	basil, saffron, ginger, mace, nutmeg, orange peel
tomatoes	basil, bay leaves, chervil, tarragon, curry powder, oregano, parsley, sage, cloves
zucchini	marjoram, mint, saffron, thyme

SEASONING BLENDS AND MIXES

Basic Vegetable Soup Blend

This blend is not only great for soups but for flavoring breads too.

1 C each chopped dehydrated carrots, onions, tomatoes and celery including tops.

1/4 C each chopped dehydrated red and green peppers and spinach

Store in container at room temperature. Makes 4 C.

Herb Seasoning Blend

This recipe is for dried herbs. Mix the herbs and crush to a coarse powder with a mortar and pestle, an electric coffee or nut mill or a food processor. The herbs should be fine enough to go through a large-holed shaker. Lemon zest is the yellow peel without the white pith. A potato peeler does the job nicely if the lemon is firm. Slice thinly and dry before adding to recipe. Lemon zest is also available in most grocery stores. This herb Blend is great on salad greens, sliced tomatoes, cucumbers, omelets, mixed with butter for a spread on breads or for vegetables.

2 bay leaves, finely chopped
4 T oregano leaves
4 T onion powder
4 tsp marjoram leaves
4 tsp basil leaves
4 tsp savory (winter savory best)
4 tsp garlic powder
2 tsp rosemary leaves
1 tsp sage leaves
1 tsp thyme leaves
1 tsp ground black pepper
1 T lemon zest

Mix together and store in labeled container. Makes 1 cup.

Salt Spice Blend

Use a pepper grinder for this salt . . Or put it through your blender till powdered if you prefer.

1/4 cup Kosher or sea salt
1 T Spike
1 tsp dill weed
1 tsp black pepper
1 tsp onion powder
1/4 tsp garlic salt
1/4 tsp curry powder
Finely grated rind of 1 lemon
1/4 tsp each: chili powder, paprika
oregano, marjoram, sage

Combine all and store in pepper grinder or put through blender using high speed. Makes 1/2 cup. Keeps indefinitely at room temp.

Spice Parisienne Blend

In a bowl, combine 1/2 C freshly ground white pepper, 1/4 C ground cloves, and 2 T freshly grated nutmeg. Store the spice in an airtight container and use it to season patties and sausage mixtures.

Mignonette Pepper Blend

In a small bowl combine equal amounts of black and white peppercorns. Toss the peppercorns, transfer to a pepper mill and grind as needed.

Spiced Pepper Blend

In a bowl, combine 1/4 C coarsely ground black pepper or mignonette, 2 T thyme, 1 T each of ground caraway seed and sweet Hungarian Paprika, and 1 tsp garlic powder. Store in tight jar and use it to season roasts and steaks.

Chili Powder Blend

In a blender combine 3 dried ancho chiles and 3-4 dried pequin chiles all with stems removed, seeded and crumbled, 1 1/2 tsp cumin seed, 1 tsp oregano, and 1/2 tsp garlic powder and pulverize the mixture. Store and use to season chili sauces.

Lemon Pepper Blend

Combine equal amounts of pepper, lemon peel, salt and chicken bouillon powder. This is great for fish or poultry.

Onion Soup Mix

1/2 C instant beef bouillon granules without MSG
1 C Instant dry minced onion
1 1/2 T onion powder

Mix together and store. Makes about 3 Cups.

Dry Onion Soup Mix

Similar to Lipton's Onion Soup Mix. Makes
1 pint = 4 envelopes
Combine the following:
7 oz Beef bouillon Powder
1/2 tsp pepper
1 cup dried minced onion
1/2 cup onion powder
1/4 cup parsley flakes
1/8 cup onion salt
To Use: 1/4 cup = 1 envelope

Make your own onion powder by blending dry onions till reduced to a fine powder. Store in labeled containers.

Curry Powder Mix

1 oz (2 T) ground coriander seed
1 T ground turmeric
1 tsp each :   ground cumin, fenugreek, ginger, allspice
1/2 tsp each: ground mace, crushed dried hot red chiles
1/4 tsp each: powdered mustard, black pepper

Blend well. Store in air-tight container. Makes 1/4 cup.

Ranch Dressing Mix

Mix together:
1/2 C black pepper (fresh ground best)
1/2 C salt
1/2 C dillweed
1/2 C garlic powder

Stir together well. Place in labeled container and store at room temperature.

Ranch Dressing
Add 2 T of the Ranch Dressing Mix to 1 C Mayonnaise, 1 C Buttermilk and 1/2 C cream cheese or sour cream.

Italian Seasoning Mix

4 T oregano leaves
4 T rosemary, crumbled
2 T marjoram, ground
2 T basil leaves
2 T thyme, whole
1/2 tsp rubbed sage
1 tsp garlic powder
1/2 - 1 tsp anise seed

Mix and crush all ingredients together. Shake before using. Makes 1 cup.

Sloppy Joe Seasoning Mix

8 T instant minced onion
3 T green pepper flakes
3 T salt
3 T cornstarch
1 T instant minced garlic
2 tsp dry mustard
2 tsp celery seeds
2 tsp chili powder

Mix together and store in labeled container. Makes about 1 cup

Mexican Seasoning Mix

10 T chili powder
2 T cumin, powdered
1 1/2 T salt
1 T oregano, powdered
1 tsp garlic powder
1/2 tsp cayenne

Mix all ingredients well. Shake before using. Yields 1 cup.

Meatloaf Seasoning Mix

13 T onion flakes
7 tsp salt
7 tsp sweet pepper flakes, powdered
3 1/2 tsp black pepper
4 1/2 tsp celery seed, powdered
2 tsp ground sage
2 tsp garlic powder
1 tsp dry mustard
1 tsp curry powder

Mix all ingredients together. Shake before using. Good in any
Wheat Meat recipe. Makes 1 cup.

Oriental Powder Mix

In a nut mill, combine 2 T each: peppercorns, fennel seed, whole cloves, Chinese cinnamon or cinnamon bark, 6 star anise (broken.)

Pulverize the mixture and sieve it into a small bowl. Store in an airtight container and use it to flavor marinades and sauces for Oriental dishes. USE SPARINGLY.

Oriental 5 Spice Mix

Blend 60 peppercorns
4 whole star anise
2 tsp fennel seed
4 (1") pieces cinnamon bark
12 whole cloves

Reduce to a fine powder and store as above. Keeps 6 months in dry place.

Chicken Seasoning Mix

A mild blend of herbs that give gluten a chicken-like flavor.
1/2 C salt
1/2 C sage
1/4 C each: ground celery seed, marjoram, thyme, and onion powder
1 T black pepper
1/2 tsp cayenne (optional)

Combine all ingredients. Store in labeled container. Makes 1 1/2 cup

Greek Seasoning Mix

2 T oregano leaves, crushed fine
1 T cumin powder
1 T sweet basil, powdered
1 T Anise seed
1 tsp garlic powder
1 tsp onion powder

Combine all ingredients well to mix thoroughly. Store in covered container in cool dry place. Makes 1/3 cup.

Seafood Seasoning Mix

A seasoning that takes on a fish flavor without being fishy. Used mostly to season gluten pieces for seafood recipes.

1/4 C lemon pepper
3 T   each onion powder, salt, and ground dill weed
2 T   each garlic salt and paprika
1 T   tsp thyme

Combine and store in container. Makes a little over 1/2 cup.
For a broth: add 1 tsp to 1/2 C water.

Tomato Seasoning Mix

If you want the tomatoes for sauces, spaghetti, stews, soups or chili, slice them and bottle using the following seasonings per quart of tomatoes being cooked.

1 T sugar or artificial - to taste
1 whole dry bay leaf
1 T whole peppercorns
2 T lemon juice
1/4 tsp onion powder

Place about 2 1/2 C tomato liquid and above ingredients in
(continued)

saucepan. Bring to rapid boil. Boil gently, covered, for 5 min. Strain and discard spices. Gently mix juice with sliced tomatoes and freeze or bottle.

Basic White Sauce Mix

Use in casseroles that call for white sauce, in creamed dishes, as a sauce for vegetables, fish, eggs, potato soup, potatoes augratin and to extend canned cream soups.

1 C flour
2 C instant dry milk or 1 1/2 C regular nonfat dry milk
1 C butter
2 tsp salt

Cut flour and dry milk into butter with pastry blender until pieces are very fine. Store in bottle in refrigerator until ready to use. Makes 5 1/2 C.
For each 1 cup medium white sauce needed, take 1/3 C mix and 1 C water. Add small amount of water to mix to blend a paste. Add remaining water and heat to boiling, stirring constantly. Boil one minute. (For 1 C thin white sauce, use 1/4 C mix; for 1 C medium, use 3/8 C mix; and for one C thick white sauce, use 1/2 C mix.) For 4 C medium white sauce, use 1 1/2 C mix to 4 C water.

White Sauce Mix with dehydrated butter

4 C non-instant powdered milk
4 C white or whole wheat flour
4 C dehydrated butter
2 tsp salt

Combine till well mixed, label container and store. Makes 2 1/2 quarts.

For a basic white sauce: Whisk together 1 C hot tap water to 1/2 C White Sauce Mix with salt and pepper to taste. Bring to boil, constantly stirring over medium heat till thick. Makes 1 cup.

Cheese Sauce Mix

2/3 C dry cheese powder (Kraft dehydrated cheese is best tasting)

6 T each non-instant powdered milk, softened butter and whole wheat flour

2 1/4 tsp onion powder

Parsley flakes (optional)

Combine dry milk, flour, butter, onion and parsley flakes, and mix until it resembles cornmeal. Keep mix tightly covered in the fridge. Makes 3/4 to 1 C.

For a basic cheese sauce: In saucepan, mix 1/2 C of the dry cheese mix with 1 C hot water. Bring to boil until it thickens. Add parsley flakes.

Guacamole Mix

2 T onion powder

1 T garlic powder

1 1/2 tsp chili powder

2 T sugar or pwd fruit sweetener

2 T sweet bell pepper flakes (softened)

2 T whole wheat flour or bean flour

1 T salt

1 T lemon powder

Mix together. Place in airtight container at room temperature. Makes 1 cup.

Note: For a more spicy flavor add: 2 tsp jalapeno powder to above mix.

Taco Mix

2 tsp instant minced onion
1/2 tsp instant minced garlic
1/4 tsp dried oregano leaves
1/2 tsp cornstarch
1/2 tsp crushed dried red pepper
1 tsp salt
1 tsp chili powder
1/2 tsp ground cumin

Mix together & store in labeled container

Spaghetti Mix

1 C dry minced onion
3 T dry minced garlic
1 C parsley flakes
1 C cornstarch
4 T Italian seasoning
2/3 C green bell pepper flakes
1 T basil
1 T oregano
1/3 C salt
1/3 C sugar

Mix together and store at room temperature. Makes about 4 1/2 cups.
Add 3-4 T of mix to 3 C tomato sauce or stewed tomatoes.

Chili Seasoning Mix

3 T flour
2 T instant minced onion
1 1/2 T chili powder
1/2 tsp Crushed dried red pepper (optional)
1/2 tsp instant minced garlic
1/2 tsp sugar
1/2 tsp ground cumin
1 tsp salt
Mix together & store in labeled container. Makes about 1 cup.

Basic Sausage Seasoning Mix

A versatile, unique blend used for a sausage flavor as well as a substitute for beef flavor.

3/4 C each: sage and salt
1/4 C each: rosemary, thyme, marjoram and basil
2 T each: cayenne and garlic powder
2 tsp black pepper

Combine together and store in labeled container. Makes 2 1/2 cups.

Herb Sausage Seasoning Mix

1 C parmesan cheese
2 1/2 T each of ground black pepper, dry crushed basil and oregano
2 T mustard seeds
1 1/2 T garlic powder
3/4 T onion powder

Place in labeled container and keep at room temperature. Makes about 2 cups.
When using to flavor commercial Gluten Flour add 1/2 C to water.

Old Home Sausage Mix

1 C salt
1 1/2 C sugar or honey powder
1/2 C green bell pepper flakes
2 1/2 T each red pepper flakes, cayenne pepper, ginger powder and ground nutmeg
2 T each thyme flakes and rubbed sage

Place in labeled container and keep at room temperature. Makes about 4 cups. When using to flavor commercial Gluten Flour add 1/2 C to water.

Pizza Pepperoni Sausage Mix

1 C each salt and sugar
1 1/2 C paprika
1/2 C each cayenne pepper, dry bell pepper flakes and dry
    minced onion
1 C anise seed
1/2 C garlic powder
2 1/2 T oregano flakes

Place in labeled container and keep at room temperature.
Makes about 7 1/2 C. When using to flavor commercial
Gluten Flour add 1/2 C to water.

Italian Sausage Seasoning Mix

1 C salt
1/2 C paprika
3/4 C fennel seed
1/2 C anise seed
1 T white pepper
1 T black pepper
4 C sugar

Place in labeled container and store at room temperature.
Makes about 7 cups. When using to flavor commercial
Gluten Flour add 1/2 C to water.

SAUCES

Saucing stands on its own as a seasoning technique. Variations on the sauce theme number in the hundreds. The following recipes are a few of the basics that you can rely on.

Hot Pepper Sauce

Similar to Tabasco Sauce.

1 tsp cayenne powder
1 T paprika
1 tsp curry
1/2 tsp chili powder
1/2 tsp white pepper
1/3 C boiling water

Combine. Cool and strain. Bottle and refrigerate. To use: Same as commercial brands. Makes 1/2 cup.

Mustard Sauce

Combine 1 C sugar, 1 C cider vinegar, 1/2 oz dry mustard, 3 large eggs in a blender or food processor and blend till smooth. Cook in double boiler, stirring constantly till thickened and smooth (approx. 10 min) or Microwave stirring frequently till thickened. Makes 2 cups.

Hot Mustard

Try it mixed with yogurt as a dip for vegetables.

1 C white wine vinegar
1 C dry mustard
2/3 C firmly packed brown sugar
2 eggs
1 tsp salt

Combine ingredients in processor or blender and mix until smooth. Transfer to the top of a double boiler and cook over gently simmering water, stirring constantly until thickened, about 5 minutes. Remove from heat and cool. Ladle into sterilized jars and cover tightly. Keep mustard refrigerated. Makes 2 cups.

BBQ Sauce

Similar to Open Pit Barbeque Sauce.

16 oz tomato sauce
1 1/2 C honey
1 C cider vinegar
1/2 C ketchup
2 tsp onion salt
2 tsp coarse grind pepper
1 tsp garlic salt
1/4 tsp cayenne pepper
1 tsp salt
3 T cornstarch
2/3 C corn oil
1/4 C dark molasses
2 tsp Postum or browned flour (page 187)
Cooks 15 minutes on medium stirring constantly. Cool and refrigerate. Makes 5 Cups.

Fry Sauce

A popular sauce for french fries and Veggie Burgers.

1/2 C mayonnaise
1/2 C ketchup
1/4 C mustard
Blend together and refrigerate.

Bearnaise Sauce

This version will make green beans or cabbage sing with flavor.

1 T chopped shallot or green onion
2 tsp minced fresh tarragon or 1/2 tsp dried
Dash pepper
Dash cayenne
1 T wine vinegar
1/4 C dry white wine (or chicken broth)
1/2 C butter or margarine
4 egg yolks *(continued)*

Mix all the ingredients, except the egg yolks, in a cup measure and heat in microwave about 2 min. until butter melts. DON'T LET IT GET TOO HOT. Beat the egg yolks in a small casserole. Gradually beat in the warm butter mixture. Start cooking for 10 second intervals, stirring well after each interval, until the sauce thickens. (1 1/2 min.) ***Serve not hot but lukewarm. If vegetables are hot they will heat sauce just enough.

Bearnaise Sauce With Tomato and Herbs

1 C béarnaise sauce
1 C peeled, seeded, and chopped tomatoes well drained
1 tsp tarragon
1 tsp chervil or marjoram
1 tsp thyme

Mix all ingredients, serve a few tablespoons for each guest with any beef steak from chuck to porterhouse. Makes 2 Cups

Hollandaise Sauce

3/4 C mayonnaise
1/2 C milk
1 T lemon juice
1/2 tsp salt
1/8 tsp pepper

Combine all ingredients in a sauce pan. Heat slowly, stirring constantly until hot. Makes 1 1/4 C.

Parsley Sauce

*Excellent On Potatoes Or With Carrots And
Vegetables With Color Contrast*

2 T butter or margarine
2 T flour
1 tsp chicken bouillon powder
1/4 tsp salt
pepper to taste
Dash of nutmeg
1 C water
2 egg yolks
1/4 cup fine-chopped parsley

Melt the butter in a saucepan. Stir in flour, bouillon powder,
salt, pepper and nutmeg. Blend in the water until smooth.
Cook, stirring several times until the sauce thickens and
boils. Continue cooking for 2-3 min. on low heat, or let sit 2
min. Bring to boil for 1 min.. Beat egg yolks in a small dish.
Blend in some of the hot sauce 1 tsp at a time. Quickly blend
the egg mixture into remaining sauce. Stir in parsley.

** For added zing, leave out nutmeg and add dash or two of
Tabasco or cayenne pepper.

Spaghetti Sauce

1 C chopped onion
1/2 C chopped green pepper (optional)
1 clove garlic, minced
1 (1 lb) can tomatoes cut up (two C or No. 2 can)
1 (15 oz) can tomato sauce (or two small cans one cup each)
1 T honey
1 1/2 tsp salt
1 tsp dried oregano leaves, crushed
1/2 tsp dried basil leaves, crushed
1/8 tsp pepper
1 bay leaf
4 C ground gluten
1 (1 lb) pkg. Spaghetti noodles
Grated Parmesan Cheese

Sauté onion, green pepper and garlic. Add tomatoes, tomato sauce, honey, salt, oregano, basil, pepper and bay leaf; cover and simmer 30 minutes. Uncover and continue cooking 15 to 20 minutes more, or until sauce is slightly thickened. Remove bay leaf. Add ground gluten and serve.

To Use: Cook spaghetti by package directions, drain and place on warm platter. Pour hot sauce over and sprinkle with Parmesan cheese, or pass cheese at the table. Makes 6 servings.

Mustard and Curry Sauce

1 C plus 2 tsp honey
2 tsp dry mustard
2 tsp curry powder
1 T white bean flour

In a bowl, combine all of the ingredients. This sauce is delicious served over rice or noodles.

Sour Cream Onion Sauce

1/2 C chopped onion
1/2 C butter or margarine
2 T flour (bean flour will do nicely)
1 C light cream
1 tsp salt
1/2 tsp white pepper
3/4 C sour cream

Cook the onion in butter until soft, but not browned. Stir in the flour, cream, salt, and pepper. Bring mixture to a boil, stirring constantly until blended. Stir in the sour cream until heated thoroughly, DO NOT BOIL. Serves 8-10. Serve on pasta, vegetables, grains.

Salsa

We use this salsa to accompany chili and cheese strata, huevos rancheros, or cheese omelets. Makes 1 quart. Combine and refrigerate up to 4 days

2 lbs fresh tomatoes, peeled and chopped or 1 1/2 qts drained canned tomatoes
1 C chopped onion
2 garlic cloves, mashed and chopped
2 T oil
2 (4 oz) cans chopped green chilies
1/2 tsp cumin
Following are optional:
1/4 to 1/3 C bean flour
1/2 C fine chopped fresh cilantro
1/2 C fine chopped green onions
1/2 C fine chopped fresh parsley
Tabasco sauce to taste

Combine all ingredients. Add Tabasco sauce to taste. Other ingredients you could add: green, red and yellow sweet peppers.

Green Pepper and Garlic Sauce

1 tsp oil
1 T chopped onion
1/2 to 1 clove garlic, minced
2 T chopped green pepper
2 C chopped fresh, or drained canned tomatoes
1/4 C bean flour
1 tsp basil
1 tsp chopped parsley
1/3 C sour cream
salt and pepper to taste
chopped green chiles to taste

Sauté onion, garlic, and pepper in oil. Add tomatoes, bean flour and spices. Cook over medium. heat until thickened. Add sour cream and chopped green chiles. NOTE: This sauce can be frozen.

Swiss Cheese Sauce

3 T butter
3 T flour
1/2 tsp salt
1/8 tsp pepper
2 C milk
3/4 C swiss cheese, shredded

Melt butter. Stir in flour salt & pepper, stirring constantly, until it bubbles. Stir in milk , continue stirring until it thickens. Stir in cheese until melted. Makes 2 1/2 C.

Herb and Butter Sauce

1/2 C butter, softened
2 T parsley, fine chopped
1 tsp marjoram
1/2 tsp thyme

Stir together and pour over baked potatoes or other vegetables.

Alfredo Sauce with Cream

This is a quick and easy sauce that goes well with any shape of pasta, especially fettuccini.

3 T butter
1 C heavy cream
1/4 C parmesan cheese (fresh grated if possible)

Cook pasta and set aside while keeping warm. Cook the butter and most of the cream in a saucepan or skillet until it begins to thicken. Add the cooked pasta, parmesan cheese and the rest of the cream. Mix until pasta is coated. Serve immediately.

Pesto Sauce

1/4 C butter, softened
14 C grated Parmesan cheese
1/2 C chopped fresh parsley
1 garlic clove, crushed
1 tsp dry basil, more if fresh is used
1/2 tsp dried marjoram leaves
1/4 C oil
1/4 C chopped pine nuts or walnuts

In a medium bowl, cream the butter with the Parmesan cheese, parsley, garlic, basil, and marjoram. Gradually add the oil, beating constantly. Add the pine nuts, mixing well. Makes 1 1/2 Cups

Serve: Toss into pasta or serve with Minestrone soup.

White Sauce

For each 1 cup medium white sauce needed, take 1/3 C mix (page 97) and 1 C water. Add small amount of water to mix to blend a paste. Add remaining water and heat to boiling, stirring constantly. Boil one minute. For 4 C medium White Sauce use 1 1/2 C basic white sauce mix (page 97) to 4 C water.

Alfredo Sauce for Pasta

Add Parmesan cheese, according to your taste, to the Basic White Sauce or Rita's White Bean Gravy and Sauce*:
Whisk 4 1/2 to 5 T white bean flour to 2 cups boiling water. Cook 3 min. over medium heat until thick. Add salt and pepper or other seasonings to taste.

*For enlightening recipes with use of beans and bean flour see "Country Beans" by Rita Bingham.

Cheese Sauce

For a basic cheese sauce: In saucepan, mix 1/2 C of the Cheese Sauce Mix (page 98) with 1 C hot water. Bring to boil until it thickens. Add parsley flakes.

Tomato Pasta Sauce

2 T olive oil
1 onion, chopped coarsely, fry until transparent and slightly browned
1 qt tomatoes
1 clove garlic, crushed
1/4 tsp Italian seasoning
1/4 tsp seasoned pepper
1/4 tsp nutmeg
1/4 tsp paprika
1 C Cooked Ground Gluten Sausage Mix (page 24)

Saute onion in oil, add tomatoes, garlic and seasonings. Simmer 15 min. Add 1 C Ground Gluten pieces. Heat through & serve over noodles or rice.

Sweet and Sour #1

1 1/2 C water
1/4 C vinegar
1/4 tsp salt
1 T catsup
1/4 C honey
1 T cornstarch
2 T water

Mix water, honey, vinegar, salt and catsup together and bring to a boil.
Slowly add cornstarch and water mixture to thicken. Green peppers, tomatoes and pineapple could be added for a variety.

Sweet and Sour #2

Mix together:
2 T brown sugar
2 G bean flour or 1 1/2 T cornstarch
Blend in:
1/4 C water
1/4 C vinegar
reserved juice from a 20 oz can of pineapple chunks
Stir in:
pineapple chunks and
2 green bell peppers

Bring to boil, simmer and stir constantly until thick (about 4 minutes.) Serve over wheat meat balls, rice or cracked wheat.

Tartar Sauce

1 C mayonnaise or cottage cheese
2 T milk or buttermilk
3 T each chopped onion, parsley, sweet pickles or sweet pickle relish

Blend mayo and milk together well and add vegetables

Marinara Sauce

Sauté:
1 T olive oil
1 clove garlic, minced
Add:
2 cans (16 oz) tomatoes
1 tsp oregano leaves
1 T dried parsley or 2 T fresh

Cover, bring to boil and simmer 25 minutes, stirring occasionally.

For a Spanish flavor add:
1 tsp cumin powder
hot sauce to taste (optional)

Lemon Sauce

Mix in pan:
1/2 C sugar
3 T white bean flour or 1 1/2 T cornstarch
Add slowly:
1 C water

Cook over medium heat, stirring constantly, until mixture thickens and comes to a boil. Boil and stir for 1 minute and remove from heat.

Stir in:
2 T butter
1 T grated lemon rind
1 T lemon juice

Serve warm or cold. Makes 1 1/4 C sauce.

Dill Sauce

Mix together:
1 C yogurt, plain
2 T mayonnaise
Add:
3/4 tsp tarragon leaves
1 1/2 tsp dill weed

Brown Gravy

2 C liquid—can be gluten water or water in which vegetables
    have been boiled (such as potatoes)
1 T Worcestershire sauce
1 tsp soy sauce
1 tsp Kitchen Bouquet
1 tsp olive oil (optional)
1 T beef or sausage seasoning base
4T flour

Mix ingredients. Heat and simmer 5 min (try adding sprigs of parsley and celery leaves during simmering process).

Chicken Gravy

1 tsp oil
4 T flour
2 T Nutritional Yeast
2 T chicken or poultry seasoning
2 C water from potatoes or vegetables

Brown flour in oil, then slowly add yeast, add juice from vegetables or potato water. (Approximately 2 C or to desired consistency.)

Mushroom Gravy - Low Fat

3/4 C unbleached or whole wheat flour
1 large onion, chopped
3 C mushrooms, chopped
1 T oil
1 C cold water
4 C boiling water
3 T soy sauce
1/2 tsp salt
pepper to taste

Place flour in skillet, brown slightly and set aside. Chop onion and mushroom in fine bits, then sauté in oil until slightly brown. Place the browned flour in 1 C cold water and blend until smooth. Pour sautéed onion and mushroom into the boiling water, bring to a boil again, then add the flour-water mixture, stirring constantly, until the flour is mixed in smoothly. Add the soy sauce and boil until the gravy thickens.

Yield: 20 to 30 servings

DRESSINGS

Dressings for salads or similar mixtures to be served as main dishes as well as side dishes, is an important flavoring factor. You may substitute sour cream or yogurt for buttermilk in your dressings & dips.

Ranch Dressing

1 C mayonnaise
1 C buttermilk
2 T finely chopped green onion, tops only
1/4 tsp onion powder
1/4 tsp garlic powder or 1 garlic clove, finely minced
2 tsp minced parsley
1/4 tsp paprika
1/8 tsp cayenne pepper
1/4 tsp salt
1/4 tsp black pepper

Combine, cover and refrigerate. Makes 2 cups.

Quick Ranch Dressing

Add 2 T of the Ranch Dressing Mix (page 93) to 1 C Mayonnaise, 1 C Buttermilk and 1/2 C sour cream or cream cheese

Garlic Dressing

1/2 C oil
1 tsp salt
3 T vinegar
1/2 C mayonnaise
1 clove garlic, crushed
1 T fine chopped capers

Place in a jar. Cover and shake until thoroughly mixed. Makes 1 cup.

Pesto Dressing

2 C fresh basil leaves
3/4 C olive oil
2 T pine nuts
2 garlic cloves, crushed
1 tsp salt
1/2 C fresh grated Parmesan cheese
Mix together using a blender. Store in container.

Green Godess Dressing

1 1/4 C homemade mayonnaise
1 C sour cream
1/2 C coarse chopped parsley
1/3 C chopped green onions
2 oz can anchovies, drained and rinsed
1 T lemon juice
3-4 T wine vinegar
generous pinch of tarragon
generous pinch of chervil (or thyme)
salt and pepper to taste

Combine all ingredients in blender or food processor and
blend until smooth. Refrigerate overnight before serving.
Makes 2 1/2 Cups.

Vegetable Marinade

1/2 C oil
1/3 to 1/2 C red or white vinegar
1/4 C minced parsley
2 garlic cloves, pressed
1 T Dijon mustard
1 tsp honey
1/2 tsp oregano
1/2 tsp basil
1/2 tsp tarragon
salt and pepper to taste
raw or slightly steamed vegetables, well chilled.
(continued)

Combine all ingredients except vegetables and blend thoroughly. Pour over vegetables and chill at least 2 hours before serving.

Honey Mustard

1/2 C Mayonnaise
1/4 C Honey
2 tsp mustard

Mix together and refrigerate. Great on salads or vegetables.

Mayonnaise

3 eggs
5 T apple cider vinegar or lemon juice
1 1/2 tsp salt
1 1/2 tsp dry mustard
3 C vegetable oil

Pour into blender the eggs, vinegar, salt and mustard and 1 C oil. Mix ingredients on low speed. Uncover and pour in slowly the rest of the oil. Makes about 1 quart.

Note: Homemade mayonnaise can be a breeding ground for Salmonella, store bought or homemade mayonnaise made with fresh uncooked eggs carry a risk when not refrigerated.

Soy Mayonnaise

1 C plain soy milk powder
1 tsp salt
1/2 tsp onion powder
juice of 1 lemon
1 C water
1/2 tsp paprika (optional)
1 C oil

Blend soy milk powder, water and seasonings. Remove cup from top of blender, gradually add the oil until the mixture thickens. Remove from blender and stir in the lemon juice.

Mayonnaise is better if let set an hour or two before serving. Yield about 3 cups.

Mayonnaise, Yolk Free

1/4 C liquid egg substitute
2 T fresh lemon juice
1/2 tsp salt
pepper to taste
1/2 tsp dry mustard
3/4 C oil

Mix together in blender the egg substitute, lemon juice, salt and pepper, dry mustard and gradually add oil while blender is on.

Pesto Mayonnaise

1/4 C liquid egg substitute
2 T fresh lemon juice
1/2 tsp dried mustard
3 T pesto
3/4 C oil
ground black pepper to taste

Mix together in blender adding oil last gradually while blender is on. Makes 1 cup. Cover and refrigerate.

Sauce Verde *(green mayonnaise)*

1 C loosely packed spinach, cleaned, dried, no stems
3 sprigs parsley
1/4 C watercress leaves
2 T lemon juice
2 T chopped green onion
1/4 tsp dry mustard
1 tsp basil or tarragon
2 anchovy fillets
1 C mayonnaise
2 T capers, drained and rinsed
(continued)

Place all ingredients except capers in blender or food processor and mix for 10-15 seconds. Transfer to a bowl and fold in capers. Refrigerate. Serve with seafood.

Tomato French Dressing #1

1/2 C condensed Tomato soup
1/2 C oil
Juice of lemon
1 T honey
1/8 tsp garlic powder
1/2 tsp salt

Blend until well mixed. Chill and serve.
Yield: approximately 1 1/2 C

Tomato French Dressing #2

1 C condensed tomato soup
1/2 C vegetable oil
1 C apple cider vinegar
2 T Worcestershire sauce
1 1/2 tsp salt
1 tsp paprika
1/2 tsp dry mustard
1/3 C honey or 1/2 C sugar
3 C vegetable oil

Put into blender all ingredients except 3 C oil. Mix on low speed. Uncover and pour in slowly the rest of the oil.

French Dressing

1/2 C fresh lemon juice
2 tsp paprika
1/2 C honey
1 tsp salt
1/2 tsp dill seed
1/4 tsp celery seed

1/4 clove of garlic
1/3 C oil

Place in blender and liquefy until all ingredients are well
mixed. Chill, shake just before serving.

Yield: approximately 1 cup

Sesame Seed Dressing

1/4 C toasted sesame seed
1/4 C lemon juice
1/4 C oil
2 T Soy Sauce

Put all in blender and whiz until thoroughly mixed.
This is delicious on any vegetable salad, but especially good
on spinach and bean sprout salad.

Thousand Island Dressing

2 C soy mayonnaise
4 dill pickles, finely chopped
2 green onions, finely chopped
4 oz can pimiento, finely chopped
4 T chopped black olives

Combine and refrigerate.

Quick Blue Cheese Dressing

1/2 C cream style cottage cheese
1 T crumbled blue cheese
1 T skim milk

Mix and refrigerate.

Tarragon Dill Mayonnaise

1 C mayonnaise
1 T finely chopped fresh tarragon
1/3 tsp dill weed

Combine all ingredients. Cover and refrigerate several hours to blend flavors. Great on breads, sandwiches or in tuna salads etc.

Buttermilk Dressing

1 quart mayonnaise
1 quart buttermilk
1 T black pepper
1 T garlic salt
1 T parsley flakes
2 T dried onion flakes
1 tsp salt

Mix in blender or electric beater. Refrigerate. Makes 2 Qts.

Basil Buttermilk Dressing

1/3 C buttermilk
1/4 C mayonnaise
1 T chopped fresh basil
1 tsp Tarragon Vinegar
1/2 tsp onion powder
1/4 tsp black pepper

Combine ingredients, mix well, cover and refrigerate. Yield 1/2 cup.

Herbed Vinegar

In a small stainless steel saucepan, heat vinegar to boiling. Meanwhile, place herb springs in clean pint jar. Carefully pour hot vinegar into jar. Cover, let stand several days at room temperature to blend flavors. If desired, strain into decorative jars and add fresh herb sprigs. Store in a cool, dark place. Yield 1 pint.

Cucumber Dressing

1 medium cucumber
1 T grated onion
1/4 C mayonnaise
2 tsp wine vinegar or lemon juice
1 T chopped parsley
2 C sour cream or yogurt
1 tsp salt
1/8 tsp black pepper

Mix in blender or with electric beaters and refrigerate.

Poppy Seed Dressing

1 1/2 tsp poppy seeds
3/4 C vinegar (apple cider is good)
1 C vegetable or olive oil
1/2 C honey
1 1/2 T grated fresh onion or minced dry onion
1 1/2 tsp salt
3/4 tsp dry mustard

Mix together and refrigerate. Makes 1 1/4 C. Shake before using.

Creamy Herb Vinaigrette

1 C plain yogurt
1/2 C mayonnaise
3 T red wine vinegar or apple cider vinegar
1 1/2 tsp honey or sugar
2 tsp Dijon mustard
1 tsp minced green onions
2 tsp fresh basil
1 tsp each of fresh chives, parsley, tarragon and thyme
1/8 tsp black pepper
1 clove garlic, crushed

(continued)

In small bowl whisk together mayonnaise, yogurt, red wine vinegar, sugar and mustard until smooth. Finely chop remaining ingredients. Mix well , cover and store in fridge. Yields 1 1/2 cup.

Mint Yogurt dressing

1/4 C plain yogurt
1/2 C mayonnaise
3 T fresh chopped mint

Combine, cover, refrigerate. Yield 3/4 C

Cream Dressing *for fruit or sweet salads*

2 eggs, beaten
2 tsp salt
3/4 tsp dry mustard
1/2 tsp paprika
1/4 C whole wheat flour
2 T honey
1/4 C each water and vinegar

Mix ingredients and cook in top of double boiler till thick. Store in fridge. When a more thin consistency is desired add juice, milk or cream before serving. Makes 1 pint.

DIPS AND SPREADS

Yogurt Honey Dip

Best fruit dip ever. A treat for the taste buds.

1 C yogurt, plain
1 C cream cheese, 8 oz pkg.
4 T frozen orange concentrate
3 T honey
4 T pineapple juice or 1 T pineapple liquid extract flavoring

Blend and refrigerate.

Vegetable Dip

1 C sour cream
1 C mayonnaise
1 T dry minced onion flakes or 1 tsp onion salt
1 T dry parsley flakes
1 tsp Bon Appetit seasoning mix (page 87)

Blend ingredients and store in refrigerator.

Fresh Herb Dip

1 C loosely packed fresh basil leaves
1/2 C slivered almonds
1/4 C loosely packed fresh parsley
1/4 C olive or Vegetable oil
1 small garlic clove, finely chopped
1 C sour cream
1/4 C grated Parmesan cheese
1 T onion soup mix (page 92)

In food processor or blender, combine basil, almonds, parsley, oil and garlic until smooth. Add sour cream, cheese and onion soup mix. Process until smooth. Chill at least 2 hours. Makes about 2 cups.

Creamy Salsa Dip

Mix salsa or picante sauce and sour cream until smooth.

Guacamole Dip

1 large ripe, mashed avocado
1/4 C mayonnaise
1 T Guacamole Mix (page 98)
1/4 C picante sauce (optional)

Spreads for Garlic Bread

Slice a French-type bread in thick slices and brush on any of the following mixtures:

#1 Melt and mix together:
1/4 C butter
3 garlic cloves, pressed
1 tsp Italian seasoning (optional)
Spread on bread & broil till browned

#2 Blend together:
1/3 C butter, softened
1/3 C mayonnaise
1/3 C grated parmesan cheese,
1 tsp garlic powder
Spread on bread & broil till browned

#3 Rub fresh cut garlic pearl on a toasted slice of bread. Brush on olive oil (optional)

Dill Herb Spread

2 C drained yogurt (hang in cheesecloth overnight, 1 cup result)
3/4 C mayonnaise
1/2 C fresh snipped chives, or 3 T dried
1 T minced onion
3 T fresh chopped parsley
2 tsp dried dill weed
1 1/2 tsp Bon Appetit seasoning mix (page 87)
2 garlic cloves, pressed
Salt and pepper to taste
Combine and refrigerate for several hours.

SOUPS

Experiment with the following herb blend and recipes to find new ways of enhancing the flavor of your soups.

Herb Seasoning for Soups

A traditional combination of herbs for seasoning most soups and bean dishes (about 3-4 C)

2 sprigs parsley
2 sprigs thyme
1 sprig marjoram
1/2 bay leaf

Tie fresh herbs together in cheese cloth when adding to recipe.

Quick Cream Soup

1 3/4 C water
2/3 C basic white sauce mix (page 97)
1-2 C cooked vegetables or combination of vegetable and
   cubed gluten wheat meat.

Blend water and basic sauce mix together in saucepan. Heat
until mixture boils and thickens, stirring once or twice. Stir in
vegetables etc. And heat through.

Cream Of Carrot Soup

4 T butter
1 large onion
1-2 carrots, sliced
1 diced potato
4 C chicken broth (page 16)
2 T minced fresh ginger
2 tsp minced fresh rosemary
1/2 C skim evaporated milk
pepper to taste
Additional rosemary sprigs, for garnish

Melt the butter in a heavy pot over low heat. Add the onion,
cover, and let it cook slowly until soft but not browned, about
30 minutes. Add remaining ingredients, except the milk &
cook until the carrots and potatoes are soft, about 30 minutes
longer. Puree in blender. Stir in the evaporated skim milk and
season to taste with pepper. Garnish with rosemary.

Potato Soup

You can easily double this recipe for more servings.

2 C cubed peeled potatoes
1/4 C chopped celery
1/4 C chopped onion
1 T parsley flakes
1/2 tsp salt
1/8 tsp pepper
1 1/2 C water
3/4 C White Sauce (page 109)
3 C milk

Combine potatoes, celery, onion, parsley, salt, pepper, and water in saucepan; cook until vegetables are tender. Meanwhile, combine White Sauce Mix (page 97) and milk in large saucepan. Cook, stirring constantly, until mixture thickens and bubbles. Add undrained vegetables and heat. Makes about 5 Cups.

Clam Corn Chowder

A rich tasting, nutritious soup that's easy and quick to make.

3/4 C White Sauce Mix (page 97)
3 C milk
2 tsp instant minced onion
1 (1 lb. 1 oz.) Can cream style corn
1 (8 oz.) Can minced clams or clam flavored gluten
1 tsp parsley flakes
1/2 tsp seasoning salt
1/8 tsp pepper

Combine White Sauce Mix, milk, and onion in saucepan; cook, stirring constantly, until mixture thickens and bubbles. Add remaining ingredients and heat. Makes about 6 Cups.

MISCELLANEOUS

Nut Milk

1 C cashew nuts
2 C water
1/2 C dates
1/8 tsp salt
1/4 grated orange rind

Blend nuts, water, dates and salt until all ingredients are liquefied. Add orange rind. If preferred, in place of the orange rind the following may be substituted for flavoring: rind from 1/4 lemon, 1 tsp vanilla or carob flavor.

Almonds, walnuts, or pecans may be substituted for the cashew nuts.

Nut Cream

1 recipe nut milk
1 banana or 1 apple

Make nut milk from above recipe, then blend with the milk, either 1 banana or 1 apple.

Sesame Milk *if you want it hot see next page*

2 C cold water in blender
2 heaping T thick sesame tahini (sesame butter)
1/2 tsp salt (scant)
1/2 tsp vanilla
Honey to taste (remember you'll be adding more water)

Blend thoroughly. Add water afterwards to make 1 to 1 1/2 quarts total, according to your taste for richness.
Delicious for use on cereals or for variation, add: 2 tsp carob powder (more or less, as your taste indicates) for delicious carob drink. See next page for hot method.

Hot Sesame Milk

If you want it hot, use boiling water, then blend. Do not heat after mixing as it separates and would need to be blended again.

Whipped Cream Substitute

1 tsp unflavored gelatin powder
3 T cold water

1/2 C additional water
1/2 C dry nonfat milk powder
1 1/4 tsp vanilla
1/4 tsp nutmeg
2 T safflower oil

Soften gelatin in the 3 T cold water and set in Pyrex custard cup placed in pan of hot water until gelatin becomes transparent. In the meantime, place the 1/2 cup water in a 6 cup mixing bowl in the freezer for 5 minutes. Also place beaters to electric mixer in the freezer for that time. Place milk powder in water in chilled bowl and beat with chilled beaters until thickened and mixture holds its shape when you lift out the beaters. On low speed beat in cooled and transparent gelatin, vanilla, nutmeg, and oil. Return to the freezer for 5 minutes. Then whip it gently with a wire whisk to keep it from separating or deflating. If it does deflate before you're ready to use it, beat it on low speed again. Makes 2 1/2 cups.

This section offers a wealth of new recipes and fresh inspiration for preparing wheat. Everything from whole wheat breads ,crackers, desserts to making wheat into a live enzyme filled food through sprouting . Enjoy wholesome eating with a flair for the gourmet.

SECTION 3

BREADS AND DESSERTS

Ingredients used in a commercial bakery just aren't as wholesome as those you use at home. Corners are cut for profit and attractive products are made to suit the public. You can use natural, wholesome ingredients instead of preservatives and chemical substitutes and have items to suit your taste.

Benefits of using Commercial Gluten Flour

Increases dough Strength
Extends shelf life
Improves texture and moisture retention in dough
Increases processing tolerance
Enhances natural flavor
Improves crumb texture

BREADS

There is nothing as satisfying as coming into a room filled with the aroma of freshly baked bread.

Hints for better bread making:
• Oil added to the dough keeps bread soft and from drying out.
• Lecithin helps preserve breads.
• Lemon juice or ascorbic acid give whole wheat bread a more tight texture and is less crumbly.
• Use a wheat flour which has a high gluten (protein) content for better rising ability.
• Adding Commercial Gluten Flour to the dough will help in the rising process.
• Be sure the yeast you are using is fresh and active. If you are not sure test some in water with honey or sugar before adding to dough. (1T yeast to 1 C warm water and 1/2 tsp Honey. Should double in size within 10 minutes). Keep yeast in freezer after opening.
• See kneading instructions in the first class on "The Quick Wholesome Foods Video".
• Brush a mixture of 1 egg white and 1 tsp water onto loaves just before baking, for a shiny golden brown crust.

- Kneading on a marble or Formica-type top requires less flour than on a bread board of wood.
- Use white whole wheat for a lighter color loaf of bread and pastries. This particular hybrid is also called "Golden '86."
- Use vegetable oil when kneading rather than flour. Too much flour can dry out the bread and make it crumbly.
- Bag and freeze bread to keep from drying out if not to be used in a few days.

Basic Whole Wheat Bread Recipe

This recipe makes 5 loaves. When cut in half, makes 3 loaves. (8 1/2 x 4 1/2 x 2 1/2) Use an electric bread mixer or knead by hand.

Mix in bowl till flour is wet:
5 1/2 C hot tap water (140° F)
1/3 C honey
5 C whole wheat flour
1/2 C Gluten Flour

Add and mix for 15 seconds:
3 T Yeast

Add:
2/3 C oil (olive or vegetable)
3 T lemon juice or 50 mg. ascorbic acid (Vit.C)
1 T salt

Add Seasonings for a variety of flavors at this point (see following recipes.)

Continue adding flour (about 7 to 9 cups) until dough forms a ball and does not stick to sides or bottom of bowl. Knead with an electric bread mixer for 6 minutes or by hand about 10 minutes (300 kneadings)

Two other methods for mixing bread are:

With oil on your hands remove the dough from bowl and form into loaves. Before placing loaf size dough into pan, fold and pound with side of fist a few times to get air bubbles out and to make a tighter and more elastic-like loaf. Make a tight ball-shape loaf and put in center of greased bread pan then place in a 125° oven or let rise on countertop till double in bulk. Turn oven up to 350° and bake for about 30 minutes or until top and bottom crust are brown. Place on wire rack to cool. Store in plastic bags.

For A Variety Of Breads

To the Basic Bread Recipe add the following ingredients after the yeast has been stirred in.

Cheddar Onion

4 T chopped dry onion
3 C grated white or regular cheddar cheese
1 T garlic powder

Cranberry Orange

Add 1/2 C additional sweetening
2 C dried cranberries
4 T grated orange peel

French, Quick

Cut oil and honey to only 2 T
Replace half the whole wheat flour with white flour, starting with the white first. For a crispy crust, glaze tops of loaves with egg white after bread has been baking about 20 min. If desired, sprinkle sesame seeds on top of glazed surface.

Garden Herb

6 T chopped stewed and drained tomatoes or any dehydrated vegetables (tomato, bell peppers, carrots, spinach, celery etc.) which have been softened a few minutes in a little hot water. 4 T each of chopped dry onions and whole dry basil. Oregano and parsley (reserve 1 T parsley to sprinkle on top of loaves before baking

Multi Grain

Replace half of the whole wheat flour with any or all of the
following ground grains.
Corn, Barley, Oat, Millet, Flax, Rye and Bran

Onion Dill

4 T dry chopped onion
1 T dill seed

Orange Date Pecan

4 T grated orange peel
2 C chopped dates
2 C chopped pecan nuts

Pepper Parmesan

1 C parmesan cheese
3 T whole dry basil leaves
1 T cracked black pepper
1 tsp cayenne pepper

Rye

Replace honey with blackstrap molasses
for dark rye
Replace half the flour with rye flour
Add 4 T caraway seeds

Spinach Feta Cheese

1 C chopped frozen spinach
1 C (8 oz.) feta cheese
2 T oregano, whole
1 T garlic powder
2 tsp black pepper

Fabulous French Bread

In large bowl, combine and mix well by hand or bread mixer.
3 C (of 6) flour
2 1/2 C hot water
3 T sugar
1 T salt
5 T oil
Stir in:
2 T yeast and remaining flour (3 C)

Dough should be barely sticky, add more flour if necessary.
Mix well. Allow dough to rest for 20 minutes and stir again.
Repeat five times for a total of 50 minutes. Turn out dough
onto oiled counter. Knead once or twice. Divide in two. Roll
each half into a 9 x 12 rectangle.
Starting at long edge, roll loosely, Seal edges. Place rolls
seam down on a baking sheet. Brush with cornmeal for a
crispy crust (optional) Cut tops diagonally three times with a
sharp knife. Bush with beaten egg white. Sprinkle with
sesame seeds. Rise 30 minutes.
Bake at 400° for 25-30 minutes. Makes 2 loaves.

Parmesan Bread Sticks

Use the above dough recipe or the Whole Wheat bread dough
(page 132) for this satisfying, popular bread treat.

Roll dough into desired size and shape, about a 1/2 x 5 inch
roll, place on baking sheet. Brush top of roll with a Garlic
Spread #2 (page 124), turn and brush on other side. Can
sprinkle extra parmesan on top. Let rise till double. Bake for
20 minutes or till browned at 350°.

French Bread with Cheese and Herbs

Make one lengthwise cut down to bottom crust, not through.
Open and spread with butter and your favorite cheese spread.
Sprinkle with sweet basil, 1/4 to 1/8 tsp rosemary, or dill weed
and a bit of parsley. Wrap in tin foil and bake 20 min at 350°.

Pizza Dough

Stir together and let sit 10 minutes:
4 C warm water
4 T each active dry yeast and sugar
Stir into water mixture and knead 10 minutes:
1/2 C oil
4 tsp salt
10 C whole wheat flour

Allow to sit for 30 minutes. Roll out 1/8 inch thick onto pans (this is enough dough for 4 pizzas). Top with pizza or spaghetti sauce, mozzarella cheese and other toppings. Bake for 20 minutes at 375°

Indian Fry Bread

A versatile flat bread used in making Navaho Tacos etc.

Mix together:
4 C flour (half white)
1/2 C non instant powdered milk
2 tsp baking powder
2 tsp sugar
Add:
1 1/2 C very hot water

Quickly work ingredients together and knead for a few minutes. Take pieces from ball of dough, pull and stretch with hands into a circle of about 6-8 inches. (or roll out on oiled counter and cut into desired shapes).
Fry in hot oil. Top with chili or refried beans, grated cheese, green onions, lettuce, thawed and slightly steamed green peas, cooked garbanzo beans, olives etc., with a drizzle of Ranch-type dressing on top.

Scones

1 qt buttermilk
8-9 C whole wheat flour or 5 cups whole wheat and 3-4 cups
    white flour
2 T active dry yeast
2 T honey or sugar
2 tsp salt
1 T baking powder
1/2 tsp soda
1/3 C oil
2 eggs

Warm buttermilk to 125°. Combine 5 cups flour, yeast and
honey in mixer bowl. Add buttermilk and mix 1 minute. Turn
off mixer. Add salt, baking powder. Soda, oil and eggs. Turn
on mixer and add remaining flour, 1 cup at a time until dough
begins to clean the sides of the bowl. Dough should be soft.

If dough gets too stiff, drizzle a little warm water over dough
as it mixes to soften. Dough may be used immediately or
covered and stored in the fridge for 2-3 weeks. Dough will
continue to rise for a while in fridge. Knead down a few
times. When ready to use, roll out room temperature dough
on a lightly floured countertop. Cut into desired shape. Let
rise. Cook on a non-stick griddle at 375° turning when brown
and cook other side or deep fry in hot oil. Makes 72 scones.

Serve hot with honey butter, jam or desired sweetener. This
bread also can be used in place of Fry Bread or tortillas.

Pita Pocket Bread

Mix together:
2 C flour
1 T active dry yeast
Add and mix well:
1 1/4 C water (120°)
1/2 tsp salt
(continued)

Gradually add another 2 cups flour until dough cleans sides of bowl. Dough should be moderately stiff. Knead 4 -5 minutes until dough is smooth and elastic-like. Do not over knead.

Form dough into 10 balls. On a floured countertop, roll each ball from the center out, into a 1/4 inch thick and 5-6 inch round shape. Make sure both sides are dusted with flour.

Place round flattened dough on a lightweight, non-stick baking sheet or on a greased 7 inch square piece of foil. Let rise 30 minutes or until slightly raised. Preheat oven to 500°.

Gently turn the rounds upside down just before placing in the oven (need not turn over when using foil.). Bake on the bottom rack of oven 5-7 minutes or till puffed and lightly browned. The instant hot heat makes the breads puff up. You may want to leave the oven door ajar slightly to make sure the heating element stays on and to let the steam out.

Note: The Pita pockets will be hard when removed from the oven and soften as they cool. While still warm, store in plastic bags or an airtight container. To serve warm, reheat in a 350° oven. Cut in half to fill with salads, vegetables, tomato and lettuce with a slice of red onion or chili, refried beans etc.

Instead of cutting in half and separating the cooked bread to fill, try folding it over a filling for a great meal.

Filling For a Pita Pocket or Whole Wheat Sandwich

Use the following vegetables combined to make a delicious wholesome sandwich:

Fresh slices of avocado, white cheese, olives, mushrooms, tomatoes, cucumbers, sprouts, purple onions.

Tortilla

Flour Tortillas can be used for a Pizza Crust

4 C whole wheat or white flour
1 T Baking powder
1 tsp salt
1/3 C oil or shortening
1 1/8 C warm water

Mix together. You may have to add a little more water when using whole wheat. If shortening is used, cut into dry ingredients. It will be a very soft dough. Knead gently just to get the dough together. Cover and let rest 5 minutes for gluten to develop. Form into balls, could use an ice cream scoop (you want a 1 1/2 inch size ball). Roll out very thin, at least 1/4 inch. Keep remaining dough covered. Bake on a hot ungreased fry pan or griddle. Brown on both sides. Makes about 24 tortillas.

Corn Tortilla

2 C corn flour (masa harina)
1 1/3 C cold water
1 tsp salt

Combine corn flour, salt and water and mix thoroughly. Cover and let rest for 20 minutes. Shape a bit of dough into a 1 1/2 inch ball. Keep remaining dough covered to prevent it from drying out. Place ball between 2 pieces of wax paper and roll out thin (or use a tortilla press). Heat ungreased griddle or pan over medium heat until a drop of water dances on the surface. Peel off wax paper and cook tortillas one at a time for 2 minutes on each side or until they turn a delicate brown. For a softer product when making soft burritos, bake less. Makes about 12 tortillas.

Filling for Tortillas or Burritos

Fill each with equal portions of Cooked Ground Gluten Pieces (sausage or taco seasoned), olives, onion, tomatoes, cheese, guacamole and sour cream. Add a taco sauce, if desired, before the guacamole and sour cream.

Bagels

Bagels can be made from virtually any yeast dough. A little butter and egg white can be added to a basic dough to accentuate the softness of the interior. Milk is used as the liquid to retard crisping of the crust during the baking.
They are always poached briefly in water before they are baked. This moistens the dough and helps them achieve their special texture.

Stir together and let stand till bubbly (about 5 minutes):
1/4 C warm water or milk
1 T active dry yeast
1 T honey or sugar

Add and mix or knead till dough forms a ball and cleans side of bowl:
1 T oil
l tsp salt
2 T butter
1 egg white
choice of seasonings or additional flavor. See recipe suggestions (next page)
2 1/2 C whole wheat flour (can use half white flour)

Let dough stand about 2 minutes. Turn on mixer again and slowly pour in about 1/2 C more water or enough to make the dough soft, smooth and satiny but not sticky. Beat another 20 seconds.

Place dough onto lightly oiled surface. Shape into a ball and cover with plastic wrap and let stand 15 minutes..

Divide dough into 12 pieces. Shape each piece into a strand about 6 inches long. Bring both ends of each strand together to form a doughnut shape. Moisten ends and pinch together to seal or roll pieces into a ball and use your finger to make and enlarge a hole in the center of each ball.

Place bagels on greased cookie sheet and let stand at room temperate another 15 minutes.

Combine in heavy pot and bring to boil:
2 quarts water
2 T sugar

Gently place bagels into boiling water. Can cook 3-4 bagels at one time. When they rise to the surface, turn them over and cook till puffy (about another 2 minutes longer. Remove bagels from water and place on greased cookie sheet.

Heat oven to 425°. Brush on egg glaze (1 egg beaten with 2 T cold water). Bake until crusts are golden brown and crisp (20 to 25 minutes). Remove from cookie sheet and cool on rack.

Flavor And Topping Varieties
Use Any Of The Following In The Basic Recipe
• 2 T instant minced dry onions
• 2 T sunflower seeds
• 1/4 C blue berries or other fruits
• 2 T dill weed and 1/4 C spinach or other vegetables
• 3 T raisins or other dried fruits
• 1 T your choice of extract flavorings.
• 1 1/2 T sugar, 1 tsp cinnamon, 1/4 tsp nutmeg, and 1/2 C raisins

Top With The Following Onto The Egg Glaze Before Baking
• Sesame, poppy or caraway seeds, cheese etc.

Quick Rolls

Use this dough for making dinner or sweet rolls as well as hamburger or hot dog buns. The texture is very light.

4 C warm water
1 C cooking oil
3/4 C honey
6 T yeast
2 1/2 tsp salt
3 eggs
15 C whole wheat flour (or until workable)

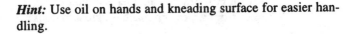

Hint: Use oil on hands and kneading surface for easier handling.

Mix water, oil, honey and yeast. Set aside 15 minutes.
Add rest of ingredients and mix well.
Form into desired shapes and let rise 20-30 minutes or until double. Bake 15 min at 400°. Makes 3 dozen.

Hamburger or Hot Dog Buns

Roll dough to 1/2 in. thick. Cut with wide-mouth canning jar ring. For hot dog bun, fold over and stretch to shape.

Orange Rolls

Replace liquid with orange juice, add 3 T dry grated orange crystals or zest. Form 3 small balls and place in cups of greased muffin tin. Spread orange glaze on warm roll.

Cheese Rolls

1 C grated sharp cheese
1/2 tsp celery seed
2 T soft butter
1 T prepared mustard
bacon bits (optional)

Mix ingredients and spread on unbuttered rolled out dough.
Roll up, slice and place in greased muffin tin.

Cinnamon Rolls

Sprinkle on rolled out and buttered dough equal parts mixed
cinnamon and date sugar.
Spread raisins and nuts if desired.
Roll up dough, slice and place in muffin tin.

Icing for Sweet Rolls

1/4 C butter
2 C sifted powder sugar
2-3 T boiling water
1/2 tsp maple extract
1/2 tsp coconut extract

Cook butter until it stops bubbling. It will be brown and
foamy. Add flavors. Beat in sugar. Add hot water for desired
consistency.

Cream Puffs

Melt and stir together:
1/2 C butter
1 C boiling water
Add and stir vigorously:
1 C whole wheat flour
1/4 tsp salt
4 eggs

Cook, stirring constantly until mixture forms a ball that does
not separate. Cool. Add eggs, one at a time, beating after
each till smooth.

Drop dough by heaping tablespoons, 3 inches apart, on a
greased cookie sheet. Bake 15 minutes at 450° then 25 minutes at 325°. Take from oven and split each one. Turn oven
off and place puffs, opened, back into oven to dry out (about

20 minutes). Cool. Fill just before serving. Fillings could be sweet (cream pie fillings, ice cream etc.) or a savory filling (vegetable or chicken salads with sprouts, tomato etc.) Makes 8 large puffs.

Whole Wheat Biscuit

Preheat oven to 400°
Mix together until crumbly:
2 1/2 C whole wheat flour
2 tsp baking powder
1/4 C brown sugar
3/4 C cold butter
Add:
1 C buttermilk or milk
Bake in greased muffin tins or drop on cookie sheet
Bake 20 minutes. Makes 12-16 biscuits.

Cheese Bread Sticks

Cut stale whole wheat bread slices into strips. Roll strips in melted butter then in Parmesan cheese. Place on cookie sheet and bake at 400° until just slightly browned. Serve with spaghetti.

Quick Croutons

Add a gourmet touch to your salad while making use of a day old bread.
Cut 3 slices of bread into 1/2 inch cubes. In medium skillet, heat 3 T olive oil and 1/4 tsp garlic powder. Add bread cubes and cook, stirring frequently, until bread is crisp and golden. Drain well on paper towels. Makes about 1 1/2 cups croutons.

Pretzels

2 1/2 C white flour
1 T active dry yeast
2 1/2 C milk

1/3 C whey
1/4 C oil
3 C whole wheat flour
1/2 C rye flour

Combine 2 cups of the white flour and the yeast. Blend milk, whey and oil. Heat until warm. Add to flour and yeast mixture. Beat about 3 minutes. Add whole wheat and rye flour plus enough of the white flour to make a stiff dough. Knead until smooth. Cover and let rise till double. Punch down, let dough rest about 10 min. Roll out and cut into strips (about 1/2 inch wide). Roll into ropes and shape into pretzels. Let raise, uncovered, 20 minutes.

Carefully lower pretzels into boiling salted water (3 T salt to 2 quarts water). Can boil 3-4 at a time for 1-2 minutes on each side.

Remove to drain on toweling, pat dry. Arrange on greased baking sheet. Brush with egg glaze (1 egg white to 1 T water) and sprinkle with sesame seeds or coarse salt. Bake 25 minutes at 350°.

Sesame Oatmeal Wheat Sticks

3 C whole wheat flour
3/4 tsp salt
2/3 C oil
3/4 C water or as needed
1 C oatmeal
2 T brown sugar
1/2 C sesame seed

Mix the oil and water, then add to other ingredients. Knead well, roll out to 1/4 inch thickness and cut into 1 1/2 inch x 1/2 inch strips.

Bake on greased cookie sheet at 350° until golden brown. Variation: Replace sesame seed with coconut and add nuts. Adjust sweetening to your taste.

CRACKERS

Now you can enjoy this popular snack with fresh flavor and for less money.

Basic Cracker Recipe

2 C whole wheat flour
1 tsp salt (omit when a recipe variation is used which calls
    for salt)
1/2 tsp each baking powder and soda
1/3 C olive or vegetable oil
3/4 C cold water
Cut flour, salt, soda & oil with fork or pastry cutter to a corn
meal texture and add water. Roll out to 1/4 inch thin. Cut
with knife or pizza cutter into squares. Bake at 350° for 10 to
12 minutes.
Seasoning Variations:
Mix following ingredients to basic cracker mix just before
the water is added

Onion Parmesan

Add:
1/2 tsp salt
1/2 tsp garlic powder
1 tsp onion salt
2 tsp spike (health store)
Sprinkle parmesan cheese on top and press into dough with
rolling pin

Spicy Cornmeal

Replace the 2 C whole wheat flour with:
1 C corn meal
1 C whole wheat flour
Seasonings:
1/2 tsp cumin, ground
1 tsp chili powder
1/2 tsp onion salt
1 1/2 tsp salt

Sprinkle parmesan cheese on top before baking.
Note: for plain cornmeal omit the cumin and chili powder.

Cheese

Add:
1 C powdered cheddar cheese
1/2 tsp salt
Sprinkle sesame seeds on top of dough before baking.

Cheddar Garlic

Add:
1 C powdered cheese
1/2 tsp garlic salt
1/2 tsp salt

Onion Dill

Add:
1/2 tsp onion salt
1/2 tsp salt
2 tsp dill weed

Sweet Cinnamon

Add:
1/2 tsp ground cinnamon
1 tsp sugar or fruit powder
Sprinkle with cinnamon sugar before baking.

Maple Pecan

Add:
1 T ground pecans
5 drops maple nut liquid flavoring
1 tsp sugar
Sprinkle with sugar before baking.

Corn Chips

2 C corn meal
1 C water
2 tsp salt
2 tsp chili powder

Mix corn meal and water with a fork. Divide in half. Add more water if necessary to hold it together. Roll out to 1/2 inch thick, between wax paper. Remove wax paper and cut into strips. Deep fry only a few seconds. Place on paper towels and sprinkle with salt, paprika or onion salt (optional)

Sprouted Wheat Crackers

This recipe comes from the Dead Sea Scrolls.

2 C sprouted wheat (about 1/4 inch long sprouts)
Preparation and Flavoring:
Grind with meat grinder or food chopper; liquefy in blender or pound with tamper on hollowed out log or rock.
Wheat, when it is sprouted, takes on a slightly sweet flavor (it becomes quite strong, the older the sprouts get). The mashed wheat can be flavored to make a sweet cracker by adding honey, brown sugar, or any sweetener. Sprinkle with cinnamon or a savory cracker can be made by adding onion, garlic or other herb salts.

Spread mashed mixture about 1/8 in. thick on well greased Teflon cookie sheet and bake in 300° oven for about 2 hrs or spread out on rock or flat wood surface in the hot sun until crunchy, about 2 to 3 hours.

Other Suggested Uses:
• Addition to bread dough (ground fine)
• A filling for egg omelets
• Warmed in butter and seasoning
• Ground with dried fruits such as raisins, dates, figs, apricots, etc., formed into balls and rolled in unsweetened coconut. Nuts can be added; also grated orange rind with a little juice to hold the balls together.

Sesame Wheat Crackers

1 1/2 C whole wheat flour
1/3 C oil
1/2 C water, or more
1 tsp salt
1/2 tsp baking powder
2 T vinegar
1 tsp caraway seeds
1/4 C sesame seeds

Mix ingredients together, stirring well. Roll dough thin. Cut various shapes with cookie cutters prick with fork. Bake on cookie sheet at 350° about 10 minutes. They over bake easily.

Corn Crackers

2 C cornmeal
3/4 tsp salt
7/8 C water
1/2 C unbleached flour
1/5 C oil
1 T sesame seeds

In large mixing bowl, mix all ingredients together well. Dough should be very hard and dry. Roll the dough out very thin, score and bake for 10-12 minutes at 350°. Watch constantly to prevent burning.
Serve hot, or store in air-tight container until used.

Wheat Thins

3/4 C water
1/3 C oil
1 3/4 C whole wheat flour
1 1/2 C unbleached flour
3/4 tsp salt

Emulsify water and oil in blender. Mix dry ingredients and oil mixture in mixing bowl. Knead as little as possible to
(continued)

make smooth dough. Roll to 1/8 inch thin on unoiled cookie sheet. Mark with knife to size crackers desired. Prick each cracker twice with fork tines.

Sprinkle lightly with salt.

Bake at 350° until crisp and very slightly browned.

Serve hot, or store in air-tight container until used.

Wheat Chips

Similar to potato chips and made from whole wheat

1 C whole wheat flour
2 C water

Mix together and season to taste with one of the following:
1/2 tsp each onion and garlic salt
1 tsp salt or vegetable salt substitute
3-4 T parmesan cheese
1 T of any seasoning in book such as taco, barbecue, onion etc.

Stir ingredients together. Pour mixture into squirt bottle, as shown in the *Quick Wholesome Foods video* (page 9), and squirt onto non-stick sprayed cookie sheet in potato chip shapes. Sprinkle with toasted sesame seeds if desired.
Bake at 350° for 10-15 minutes or until crisp. Check occasionally and turn chips over if middle is not cooking as fast as the outside.
Note: the thinner the batter, the more crisp the chips.

For cold cereal flakes: season batter only with salt to taste and a little sweetening if desired. Bake as above for Wheat Chips.

Potato Wheat Chips

1/2 C whole wheat flour
2 C Potato Buds
1 T salt
6 T corn oil (not shortening)
10 T cold water

Mix potato buds with flour and salt. Combine oil and water. Add to dry mixture until you can shape it with your hands into stiff ball of dough.
Roll out paper thin between 2 sheets of waxed paper. Invert onto a lightly greased cookie sheet. Carefully remove top sheet of wax paper, allowing dough to bake on the remaining sheet supported by the greased cookie sheet.
Bake at 350° for 15 minutes. Cool 5 minutes and remove the paper. Break into pieces. Salt to taste. Store in paper bags, to keep crisp.

QUICK BREADS
Muffins, Cakes, Cookies, Pancakes, and Waffles

BASIC BRAN BATTER
A batter without yeast, Ready to Bake

This innovative Quick Bread recipe is the answer for cooking wholesome food for the 90's! Quick, easy, inexpensive, and a hit with any age. Always keep a supply in the refrigerator for instant nourishment. Because of the acid in the buttermilk, this batter will keep 5 weeks or more. You add only a few extra ingredients to this basic batter to enjoy a variety of muffins, cakes, cookies and basic pancakes or waffles. For muffins, cakes and cookies use the batter plain, Create your own variations or follow recipes in this section.

Mix and Set Aside to cool:
4 C wheat cereal flakes*
1 C 100% Bran
2 C boiling water

Cream together:
1 C vegetable oil
2 1/2 C honey or desired sweetener
4 eggs

Add and Mix well:
2 T soda
2 tsp salt
4 C buttermilk
cooled bran and water mix
7 C whole wheat flour

Store in tightly covered container and keep refrigerated at all times.

Yield: 13 Cups batter

* The Bran called for here can be a variety of any Bran
Combinations:
1. 4 C Bran cereal buds or 100% All Bran.
2. 1 C cereal flakes (wheat, oats, corn, rice) and 1 C unre-
    fined bran
3. 1 C thick raw bran
(which has been reserved from gluten making)
Note: water which rises to the top of settled bran, should be
drained off, so you have pure bran. See page 13.

Tips and Hints:
Yield when using 3 Cups Basic Batter Mix:

12-16 muffins
1 cake
24 cookies

Baking Times:
Vary according to altitude, your oven, moistness of batter.

The following are Temperature suggestions:
Muffins = 375°-400° for 20-25 minutes
Cookies = 375° for 10-15 minutes
Cakes = 375° for 30-45 minutes
Pancakes = medium to medium-low heat
Waffles = regular waffle heat

Temperature:
375°-400°
Preheat oven. Once you determine the temperature that works
best for you, that temperature should be about the same for
any variety of muffin.

MUFFINS

Muffins are for all occasions and compliment almost any meal. Great for breakfast, snacks, lunch, and dessert. Serve with fruit in the morning or bag and eat on-the-run for a wholesome start and energy booster. Take extra for a mid-morning snack, to round out a light lunch or for a satisfying but a not-too-sweet ending to a meal.

Tips:

To Adjust Batter for Perfect Muffins:
Depending on the Bran used in the basic batter or added ingredients, you may need to add liquid to thin down batter or flour to thicken.
To Thin- add any of the following, 1 tablespoon at a time: water, milk, yogurt, or fruit juice.
To Thicken- add a little at a time:
flour of any kind, rolled oats, wheat germ, or corn meal.

To Sweeten:
Some recipes will indicate additional sweetening.
The following suggestions for natural sweeteners may be useful when you still need a sweeter batter and do not want to use white sugar. (Check health food stores for more.)
Honey, pure maple syrup, rice syrup, fruit syrup, molasses, sorghum, barley malt, Sucanat, turbinado sugar and Stevia (an herb blend).

Prepare Tins:
According to the additional ingredients, the yield for 3 cups batter will be from 12 muffins. Grease, use paper liners, or cooking spray in muffin tins.

Baking Times:
Depending on the moistness of the batter or type of muffin, baking time should take 20 to 25 minutes, baking at 350° to 375°.

Be Creative:
Interchange flavors and add additional ingredients to suit the desired taste.

RECIPES FOR SWEET MUFFINS

Use this batter plain for a simple, quick muffin or follow recipes using **3 Cups** of the Basic Batter. Yield 12-16 regular size muffins.

Apple Cinnamon

1 C finely chopped, peeled apples
3 tsp cinnamon
1/2 tsp each, nutmeg and cloves
1 C each, raisins, nuts (optional)

Banana Nut

2 well mashed or blended bananas
1 T vanilla
1 C chopped walnuts
Other options- add any one of the following:
Peanut butter, corn meal, poppy seed, raisins, cranberries, figs, prunes, dates, or other dried fruits; chocolate chips, pecans, butterscotch or peanut butter chips; Fresh or frozen berries.

Black Cherry

1 C fresh or (drained) canned black cherries. TOSS with 1 T
  flour before adding to batter.
1 T cherry extract or vanilla

Blueberry

1 C blueberries-fresh, frozen, or canned.(drain well).
Add just before baking.
1 T cinnamon or vanilla
2 T grated lemon peel (optional)

Brown Sugar Butter

For a richer, sweeter plain muffin, add:
1/2 C additional brown sugar
6 tsp melted butter
1 T vanilla

Carrot

1/2 C raw and finely grated carrot
1/4 C pure maple syrup, honey or molasses
2 tsp ground cinnamon or orange extract
2 T grated orange peel
1/2 to 1 C each raisins, nuts, coconut (optional)
Topping-cream cheese (optional)

Chocolate

4 tsp baking cocoa
1/4 C brown sugar or white sugar
1/2 C raisins and or nuts (optional)

Coconut

1/2 C chocolate or carob chips
1 C instant potato buds
1 T coconut extract

Coconut Mandarin

1 C instant potato buds
1/2 C coconut
1/2 C mandarin oranges
1/2 C chopped nuts

Date Nut

> 1 C chopped dates
> 3/4 C chopped nuts
> 1 tsp vanilla

Gingerbread

> 1 T cinnamon
> 1 tsp each, ginger and cloves
> 1/2 C molasses (less if using black strap)

Honey Lemon

> 1 T grated lemon peel
> 2 tsp lemon juice or 1 T lemon extract
> 1/2 C honey
> Topping-lemon glaze (optional)

Jam Filled

> 1-2 tsp any jam flavor (preserves).
> Fill muffin tin 1/2 full of basic muffin mix.
> Drop preserves into batter center of each cup.
> Add more muffin mix to fill cups 3/4 full.

Oatmeal

> 1 C rolled oats
> 2 tsp cinnamon
> 1 tsp nutmeg
> 1 C each, raisins and walnuts (optional)

Oatmeal Lemon Apple

> *Add to Oatmeal Muffin recipe:*
> 1 C finely chopped apples
> 2 T lemon juice from concentrate

Orange

1 T grated orange peel
2 T orange juice or 1 T orange extract
3/4 C chopped nuts (optional)

Orange Cranberry

Add to above recipe:
1 C dried or fresh cranberries
1 T each finely grated orange peel and orange extract

Peanut Butter

1 C softened creamy or crunchy peanut butter
1/2 C brown sugar or honey
1/2 C chopped roasted, salted peanuts (optional)

Peanut Butter Chocolate

Add to above recipe:
2 T baking cocoa

Poppy Seed Variations

2 T poppy seeds and any one of the following flavorings:
    1 T almond extract
    1 T each, grated lemon peel and lemon extract
    1 T rum extract

Topping Suggestions: Use a glaze with same flavor used in the muffin recipe.

Pumpkin

1 C canned pumpkin
1 T cinnamon
1 tsp each, nutmeg, ginger, and cloves
1/2 C chocolate chips (optional)

Prune

1 C chopped dried prunes
3 T lemon peel

Raspberry Lemon

1 C fresh or (drained) canned raspberries tossed w/ 1 T flour
2 tsp vanilla or raspberry extract
1 T grated lemon peel or 1/2 C lemon yogurt
1/2 C chopped pecans (optional)
Top with lemon glaze and place raspberry on top.

Tropical Fruit

1/2 C dried fruit bits (papaya, pineapple, banana chips, etc.)
1/4 C flaked coconut
1/2 C chopped macadamia nuts
1-2 tsp grated lemon peel

CAKES
Using any of the flavors for sweet muffins

Use 6 C Basic Bran Batter (page 152) for a 9x13 cake pan
3 C for 9" cake pan, or 5x9 pan
1 C for a 3x5 pan.

Put batter in greased and floured pan.
Bake at 350° to 375°, 30 to 60 minutes or until done, unless otherwise indicated.

Test for doneness: When top springs back when lightly touched. Cool 10 minutes and remove from pan.

TOPPINGS

These can be helpful for special occasions or to encourage those who tend to shy away from a more dense bran muffin.
When sugar is called for it can be replaced with a sweetening of your choice.

GLAZES

Basic recipe:
In mixing bowl combine 1 1/4 C confectioners sugar to 1/4 C juice. Stir till smooth.(for smaller amount: combine 1/2 C confectioners' sugar to 1 T juice)

Recipe suggestions:
> ### Citrus glaze
> 1 1/4 C confectioners' sugar
> 1/4   C orange or lemon juice
> 1 tsp citrus peel (optional)
> 1 tsp vanilla
>
> Mix until smooth and spread on warm muffins, cookies or cakes.

Butterscotch

Melt- 1/2 C butter
Add:
1 C brown sugar
Blend in:
1/3 C evaporated milk
1 C coconut
1 C nuts
1 tsp vanilla
Spread on hot muffins, cake, or cookies.
(This is especially good with oatmeal recipes).

Carob Frosting

Cream together:
2 T softened butter
1/4 C honey
1/4 C cream or canned evaporated milk
1 tsp vanilla
Add and beat until smooth:
2/3 C powdered non-instant milk
1/3 C carob powder
Note: 1 tsp or more hot water may be needed to get it
smooth.

Chocolate

Mix and beat until smooth:
4 T cocoa
3 C powdered sugar
4 T softened butter
2-3 T milk
1 tsp vanilla
Variation: 1 tsp grated orange peel

Chocolate Nut

Mix and bring to boil:
1/2 C melted butter
4 T baking cocoa
1 tsp vanilla
6 T milk
Remove from heat and add:
4 C powdered sugar
1 C chopped nuts
Frost while muffins, cake, or cookies are hot.

Cinnamon Sugar

Sprinkle over muffins or cookies before baking:
1 – 1 T granulated sugar mixed with 1/4 tsp cinnamon
2– 2 T granulated sugar mixed with 2 tsp cinnamon

Cinnamon Nutmeg

1/3 C sugar
1/2 tsp each cinnamon and nutmeg
Mix and sprinkle on before baking

Cream Cheese

Cream together:
8 oz. softened cream cheese
1/4 C softened butter
1 1/2 C powdered sugar

Creamy Orange

Combine and mix well:
1 (8 oz.) pkg. softened cream cheese
1 T sugar or honey
1 T each orange juice and finely grated peel
Cover and chill

Easy Fudge Frosting

Cream together:
3/4 C honey
1/4 C butter
1/4 C evaporated milk
1 C (6 oz) semi-sweet chocolate pieces
1 C marshmallow cream
1 tsp vanilla

Lemon Struesel

Combine in bowl until crumbly and sprinkle over batter before baking:
1/4 C flour
1 tsp cinnamon
1/2 C brown sugar
1 tsp grated lemon peel
1/2 C chopped pecans
2 T melted butter

Nut Cream

1/2 C melted butter
1 C brown sugar or maple syrup
1/3 C evaporated canned milk
1 tsp vanilla
1 C each coconut and chopped nuts
Melt butter. Add sweetening and remaining ingredients.
Spread on warm muffins, cake, or cookies

Orange Sugar

1 Mix until crumbly and sprinkle over batter before baking:
1/4 C sugar
1 tsp grated orange peel
2 Remove hot muffins from tin and dip top into:
1/4 C melted butter, then in a mixture of:
1/4 C sugar
1 tsp grated orange peel

Peanut Butter Chocolate

1/2 C peanut butter
1/2 C cream
1/3 C cocoa
1 tsp vanilla
1/4 tsp salt
3 C powdered sugar or more if needed
Beat together until creamy.

Pecan Streusel

Combine:
1/4 C each chopped pecans, brown sugar, and flour
Stir in until mixture resembles moist crumbs:
2 T melted butter

Powdered Sugar

Dust with powdered sugar after hot muffins or cake have
cooled about 5 minutes.

RECIPES FOR SAVORY MUFFINS

Adjust the Basic Batter Mix for this type muffin by reducing sweetening to
only 1 cup and add 1 more tsp salt. Same cooking time. Use any of the following recipes to 3 cups batter:

Caraway

2 tsp caraway seed
1/2 C shredded cheese
Half way through baking time, sprinkle cheese on top.

Caraway Dill

1 T caraway seed
2 tsp dill weed

Chili Taco

6 T diced green chilies
1 C sliced olives
3/4 C corn meal
3 T taco seasoning OR
2 tsp chili powder
1/2 tsp ground paprika, cumin, oregano

Cornbread Pumpkin

1 C corn meal
1 C canned pumpkin
1 can (4 oz.) chopped green chilies (optional)

Dill Cheese

4 tsp dill weed
1 T onion salt
1 tsp fresh ground black pepper
1 C grated sharp cheddar cheese
TOP with grated cheese after 20 minutes of baking.
Bake another 5 minutes.

Garden Mix

1 C combination (or only one) of
shredded zucchini, yellow squash, carrots

Herb Bacon

4 T bacon bits (made from soy)
4 tsp basil leaves
2 tsp each marjoram, oregano, thyme leaves

Herb Parmesan

1 C fresh chopped parsley, basil or cilantro
2 T olive oil
1 C ground parmesan cheese
1/2 tsp each garlic and onion salt
salt and pepper to taste (optional)

Sausage Corn

1 tsp sausage seasoning
3/4 C corn meal
Serve with honey butter

PANCAKES AND WAFFLES
MADE WITH THE BASIC BRAN BATTER
The following recipes can be used for both pancakes & waffles

Use either the Sweet or Savory Basic Batter.

To 1 1/2 C Basic Bran Batter (page 152) add:
1/2 C milk or buttermilk
1/2 C flour
1 egg
Cook at medium to medium-low heat for a nice golden
brown look.
Yield: 6-7, 6 inch pancakes

For lighter Waffles:
To 1 1/2 C Basic Batter add:
3 tsp baking powder
4 T water or milk

PANCAKES AND WAFFLES

NOT USING THE BASIC BRAN BATTER

Regular Pancake Recipe

1 1/4 C whole wheat flour
2 tsp baking powder
1/2 tsp salt
1 egg, beaten
1 C milk
2 T oil

Wheat Kernel Pancakes

Blend in liquefier, 4 min:
1 C whole kernel wheat (not cooked)
1 C milk
Add and blend until mixed well:
3 eggs
1/4 C oil
2 tsp baking powder
1 tsp soda
1 T honey or sugar
1/4 tsp salt
Fry on griddle or heavy fry pan

Pancakes (egg and dairy free)

2 C whole wheat flour
2 T baking powder
1 C soy milk
1 T vinegar
honey and salt to taste
Combine and fry on medium heat.

Note: The pancakes will be tough if batter is saved and used
the next day.

Variations:
Blueberry:
When underside is browned, sprinkle drained blueberries
over each pancake. Turn and brown top side.
Apple:
Add to batter 1/2 C finely chopped apple
1/4 tsp cinnamon

Waffles

2 eggs (separated-optional)
1-1/4 milk (when using buttermilk, mix with 1/2 tsp. soda)
1/4 C oil
1/2 tsp salt
3 tsp baking powder
1 3/4 C whole wheat flour (could mix half white)

Mix all moist ingredients with electric beaters. Add dry ingredients. Stir well but don't over mix. Bake in preheated waffle
iron. For a light and crisp texture,separate the eggs and beat.
Fold in stiff whites just before baking.

TOPPINGS FOR PANCAKES AND WAFFLES

Cranberry-Orange Butter

Mix:
1 C whole cranberry sauce (canned)
1/2 C honey
2 tsp orange rind

Maple Syrup

#1 Combine:
1/2 C honey
1/2 C brown sugar
1/2 C water
Heat until sugar dissolves.
Add a few drops of maple flavoring and 1 T butter

#2
4 C water
2 C brown or raw sugar
1/2 C corn syrup or 1/4 C honey
2 tsp maple flavoring

Stir and heat. (Simmer in saucepan for 1 hr for a thicker syrup.)

Whipped Butter

Cream 1/2 C butter with mixer or wooden spoon until fluffy

Honey Butter

1/2 C softened butter
3/4 C honey

Cream or whip butter till fluffy, using an electric mixer or wooden spoon. Gradually add honey beating till smooth.

Variations:
• 2 tsp grated orange or lemon peel
• a flavor of choice from oils or extracts

QUICK BREADS *NOT USING THE BASIC BRAN BATTER*

Banana Nut Bread

1/2 C butter or margarine
1 C sugar
2 eggs, beaten
3 bananas, mashed
2 C fine whole wheat flour
1 tsp baking powder
1 tsp soda
1 C nuts *(continued)*

Cream butter and sugar. Add eggs, then mashed bananas. Sift dry ingredients and add to first mixture. Add chopped walnuts. Bake in greased 9 x 5 loaf pan at 350° for 1 hr. When baked in small pans, bake 40 min.

Pumpkin Bread

Blend together:
3 1/3 C whole wheat flour
2 1/2 tsp salt
1 T Pumpkin spice (page 84) or 1 tsp cinnamon and 1/2 tsp each cloves, ginger, allspice and nutmeg.
Add to the following mixture and mix well:
2 C cooked pumpkin (or canned) mixed with 2 tsp soda
2 2/3 C brown sugar or honey
4 eggs
2/3 C oil
1 1/2 tsp salt
1 C each raisins and chopped nuts (optional)

Batter should be the consistency of banana bread. If the pumpkin is extra dry, add water. Bake 40 to 45 minutes at 350 degrees in 2 greased 5"x9" loaf pans.

Rhubarb Bread

1 1/2 C brown sugar
1 egg
2/3 C oil
1 C buttermilk
1 tsp vanilla
2 1/2 C fine whole wheat flour
2 tsp salt
1 tsp soda
2 C raw cut rhubarb
1/2 C nuts
Topping:
1/2 C raw sugar
1 T soft butter or margarine
1/2 tsp cinnamon

Mix together brown sugar, egg, oil, buttermilk and vanilla. Sift together and add flour, salt, and soda. Fold in rhubarb and nuts. Pour into two greased loaf pans. Combine topping ingredients and top. Bake at 350° for 60 minutes. Freezes and stores well.

Carrot Cake

1 1/2 C cooking oil
4 eggs
1 C raw bran
1 1/2 C raw sugar
1/2 C honey
3 C whole wheat flour
1/2 tsp allspice
2 tsp cinnamon
1/2 tsp salt
2 tsp soda
5 C finely grated carrots (more if using pulp from carrot juice)

Combine and mix thoroughly (about 1 min on high speed with electric beaters), oil, eggs, bran, sugar, honey.
Sift in flour, spices salt and soda.
Blend in grated carrot. Pour into greased 8" x 12" cake pan and bake at 350° for 40-45 min.

Icing #1

4 oz cream cheese
1/2 stick butter
1 tsp vanilla
1 tsp maple flavoring
1 C powder sugar
1 C powder milk (non-instant)
1/4 C honey
add halved walnuts and coconut

Mix all together. Spread on cooled cake (add a little milk if necessary to spread easier).

Icing #2

2/3 C brown or raw sugar
1/4 C melted butter or margarine
1 C Gluten Crunch (page 37)

Mix butter, sugar, and crunch (can add coconut). Spread over top of hot cake. Place under broiler until icing bubbles and is slightly brown, 2-3 min.

Anything Goes Cake

To this basic cake recipe, just about any leftovers can be added. Try tomato soup, mashed potatoes, vegetables or fruit, squash, etc.

3 C fine whole wheat flour
2 C brown sugar or 1 1/2 C honey
4 T carob powder or cocoa
2 tsp soda
1 tsp salt
3/4 C cooking oil
2 T vinegar
2 tsp vanilla
2 C liquid, mashed fruit or vegetable

Mix dry ingredients, then add liquids. Mix together. Bake in ungreased 9" x 13" pan at 350° for 30 min.

Boiled Raisin Cake

Boil for 5 min and cool
1 3/4 C honey
2/3 C oil
2 C raisins
3 C water

Mix together:
4 C fine whole wheat flour
2 tsp baking powder
2 tsp soda
2 tsp cinnamon
1 tsp nutmeg
1 tsp salt

Add :
nuts
carob or chocolate chips
coconut
dates (omit nutmeg and cinnamon)

Mix together and pour into greased 9x13 cake pan. Bake 1 hr at 350°.

Whole Wheat Sponge Cake

(like Angel Food cake)
6-8 eggs (depending on size), separated
1 1/2 C raw or brown sugar
1/2 C water
1/2 tsp vanilla
1/2 tsp lemon extract
1/4 tsp almond extract
1 1/2 C sifted whole wheat flour
1/4 tsp salt
1 tsp cream of tartar

Beat with mixer, yolks, sugar, water and flavorings 5 to 7 minutes. Sift flour and salt 3 times. Add to egg mixture while continuing to
(continued)

beat. Beat egg whites and cream of tartar until very stiff, 5 minutes.
Fold immediately into first mixture. Bake in ungreased angel
food pan for 1 hour at 350°. Invert pan and allow to cool 2
hours before removing. An excellent cake for fruit toppings,
lemon sauce and whipped cream.
(This is one recipe that must be followed exactly as written).

Honey Orange Sponge Cake

Use recipe above for Whole Wheat Sponge Cake.
Replace sugar for honey. and the lemon and almond extract
with 1/3 C orange juice and 1/2 tsp powdered cloves.

COOKIES
MADE FROM THE BASIC BRAN BATTER MIX (PAGE 152)

Use the same quantity and variety of flavors as found in the
muffin recipes. The only difference is to drop the dough by
spoonful onto a greased cookie sheet and bake at 375° for 10
-15 minutes.

A basic rule to produce a more firm cookie is to add any of the
following flours of various kinds, rolled oats, wheat or corn flakes.

To 3 cups of the Basic Bran Batter use any of the muffin
recipes starting on page 155:
1 C additional flour or
1 1/2 C rolled oats

COOKIES NOT MADE FROM THE BASIC BRAN BATTER MIX

A Basic Cookie Recipe

Mix:
1 egg
1/2 C cooking oil
1/2 C butter
1 1/2 C brown sugar
1/2 C honey

Add:
1/2 C water
1 tsp vanilla
1 tsp cinnamon
1/2 tsp salt
1 tsp soda
Add:
2 C whole wheat flour
1 C white flour
2 C quick oats

Bake 13-15 min at 350°
Variations:
Add to this basic recipe any one of the following:
1 C raisins, dates or other dried fruit
1 C chopped nuts
1 C carob or chocolate chips
2 C cooked mashed carrots; replace water with orange juice
1 C peanut butter, accompanied with 2 T Sour cream
1 C pumpkin with 1/2 tsp each nutmeg, ginger and cloves
1 C coconut

Carob Brownies with Bran

Fudge-Type
1/2 C butter
1/2 C raw bran
1/2 C raw sugar or honey
1 tsp vanilla
2 eggs
3/4 C dipping carob, melted (3/4 lb)
2 T carob powder
3/4 C whole wheat flour
1/2 C chopped walnuts

Cake-Type
Add to above mixture:
1/2 tsp baking powder
1/4 C raw bran
1/4 C flour
(continued)

Cream butter, sugar or honey, and vanilla. Beat in eggs.
Blend in carob (which has been melted in double boiler). Stir
in remaining ingredients.
Bake in wax-paper lined 8 x 8 x 2 inch pan 30 to 35 min at
325°. Cool. Cut into squares.

Carob Frosting #1

Cream 2 T butter with 2/3 C powder non-instant milk. Add
1/3 C carob powder. Mix well. Add 1/4 cup honey, 1/4 C raw
sugar, 1/4 C cream and 1 tsp vanilla. Beat till smooth. Spread
on cool cake.

Carob Frosting #2

Melt 1/2 C dipping carob over hot water. Cool to lukewarm.
Beat 2 egg whites till fairly stiff. Add 1/3 C honey gradually;
continue to beat. Then add carob and 1/2 tsp vanilla. Beat till
good consistency to spread.

Whole Wheat Brownies

Blend together:
3/4 C baking cocoa
1/2 tsp baking soda
1/3 C oil
1/2 C boiling water
Stir in:
1 1/2 C honey
1/2 C brown sugar
2 eggs
1/3 C oil
Add and mix together:
1 1/3 C flour
1 tsp vanilla
1/4 tsp salt
1/2 C coarsely chopped nuts
2 C chocolate or carob chips

Pour into a greased 13 x 9 baking pan. Bake at 350° for 35-
40 minutes or until brownies begin to pull away from sides of
pan. Cool. Makes 3 dozen.

Applesauce Oatmeal Cookies

2 C applesauce
2 tsp soda
1 C shortening
1/2 C honey or 1 C sugar
3 eggs
4 C whole wheat flour
2 tsp soda
1 tsp each of salt, cloves, cinnamon and nutmeg
1 C rolled oats
1 C nuts
1 C raisins (optional)

Mix applesauce and soda together. Set aside till bubbles
appear. In mixing bowl cream shortening, sugar and eggs.
Add applesauce. Add dry ingredients. Drop on greased cookie sheet. Bake at 350° for 12 to 15 min.

Sugarless Coconut Date Cookies

1 C dates, chopped
1/2 C raisins
1/4 C water
1 1/4 C quick oats
1 1/3 C unsweetened coconut
2 T oil
1 tsp vanilla
1 tsp cinnamon or cardamom
1/4 tsp salt
6 C shredded apples

Let dates and raisins soak in 1/4 C water about 15 min. Mix
other ingredients, except the apples. Stir this mixture with the
dates, raisins and apples. Drop on greased cookie sheet. Bake
for 20 min at 375-400°. This is a very moist cookie.

Sugarless Date And Nut Cookie *(frozen dough)*

1/2 C butter
1 egg
2 tsp vanilla
1 C whole wheat flour
1 tsp baking powder
1/4 tsp salt
1 C each chopped pitted dates, shredded coconut and
    chopped walnuts.

Beat butter, egg and vanilla with an electric mixer until
smooth. In a separate bowl, combine flour, baking powder
and salt. Gradually add flour mixture to creamed mixture;
beat until blended. Stir in dates, coconut and walnuts.
Form dough into two 1 1/2 inch rolls and wrap in wax paper
or foil. Chill in freezer until firm enough to slice easily
(about 2 hours). Can freeze for as long as 1 month. Bake at
350° for about 12 minutes or till brown. Makes 5 dozen.

Filled Cookies

Filling:
3/4 C water
2 C dates, raisins or any softened, chopped, dried fruit
1/2 C honey or 3/4 C raw sugar
1 C nuts, chopped

Mix water, dates and honey. Cook on stove 5 min, stirring all
the time. Then add the nuts and let cool.

Cookie Dough:
1 C vegetable oil
1 C each raw sugar and honey
2 eggs
1/2 C water
1 tsp vanilla
4 C whole wheat flour
1/2 tsp salt
1 tsp soda
1 tsp cinnamon

Cream: Oil, sugar, honey, eggs
Add water and vanilla
Mix ingredients together and add to creamed mixture
Place 1 tsp dough on cookie sheet
Add 1 tsp filling on top of this, then another tsp of dough to
cover filling. Place 2 inches apart on greased baking sheet.
Bake 15 min at 375°.

Couch Potato Cookies

Beat together:
1 C butter
1 C sugar
1 egg
Add:
2 tsp coconut extract
1 tsp soda
1 3/4 C whole wheat flour
1 1/2 C potato flakes or buds
Drop by teaspoons onto greased cookie sheet. Bake at 375°
for 10-12 minutes.

Soft Sugar Cookie

Mix together and set aside:
1 C commercial sour cream
1 tsp vinegar
In mixing bowl blend together:
2 eggs
3/4 C butter
1 3/4 C sugar
1 1/2 tsp vanilla
2 tsp soda
1/2 tsp salt
1/8 tsp lemon juice (optional)
Add sour cream mixture.
4-6 C flour (add till workable)
Roll out and cut with cookie cutter. For sugared top, pat one
side on plate of sugar. Bake at 375° on greased cookie sheet
for 8 minutes. Serve plain or spread glaze on warm cookie .

Chocolate Chip Cookies

1/2 C vegetable oil
3/4 C honey
1 egg, beaten
1/2 tsp vanilla
1 C whole wheat flour
1/2 tsp each of soda and salt
1/2 C nuts
1/2 C carob or chocolate chips

Blend oil, honey , egg and vanilla. Add flour soda & salt.
Then add nuts and chocolate or carob chips. Drop by spoon-
fuls on greased cookie sheet.
Bake at 375° for 10 minutes.

Orange Walnut Cookies

Beat together:
3/4 C honey
3/4 C vegetable oil
1 egg
Add and mix well:
1 1/2 C cooked, mashed carrots
1 tsp vanilla
1 tsp baking powder
1/2 tsp salt
2 C whole wheat flour
1 C chopped walnuts

Drop by teaspoonfuls on greased cookie sheet. Bake 15 minutes at 375°.

Glaze: Grated rind of 1/2 orange, pinch of salt, and juice of
1/2 orange. Add powdered sugar to a spreading consistency.
Spread over warm cookies.

Peanut Butter Cookies

1/2 C vegetable oil
1 C brown sugar or honey
1/2 C peanut butter
1 egg, beaten
1 T sour cream
1 tsp soda
1 1/2 C whole wheat flour

Cream oil and honey or sugar. Add peanut butter, egg and mix well. Add sour cream, soda & flour and beat well. Form into balls, the size of marbles and place on greased cooking sheet. Press with a fork. Bake at 350° for about 10 minutes.

Peanut Butter Almond Cookies

2 C whole wheat flour
1 tsp baking powder
3/4 tsp salt
1 tsp cinnamon
2 eggs
1/4 C milk
1/2 tsp vanilla
1/2 C each peanut butter, honey and brown sugar
1/2 C butter or vegetable oil
1/2 C chocolate or carob baking chips
1/2 C chopped toasted almonds, cashews or walnuts
1/2 C each flaked coconut, raisins and granola-type cereal

Combine flour, baking powder, salt and cinnamon and set aside. In large bowl, place eggs, milk and vanilla. Beat with an electric mixer. Beat in peanut butter, honey, brown sugar and butter until creamy.

Stir in flour mixture until blended, then stir in chocolate chips, nuts, raisins, coconut and granola until blended.
Drop batter, 1 heaping teaspoon at a time, about 1 inch apart on a lightly greased cookie sheet.
Bake at 375° for 10 minutes or until cookies are golden on the bottom. Cool on rack. Makes 5 dozen cookies.

Date Squares

2 C pitted chopped dates
2 C water
1 tsp lemon juice
1 tsp grated lemon rind
1 tsp vanilla
1/8 tsp salt
1 1/2 T cornstarch and 1/3 C water

Mix dates and water together in a 2 quart saucepan. Cook slowly until dates become soft. Add lemon juice, lemon rind, vanilla and salt. Mix the cornstarch with remaining 1/3 C water until dissolved. Pour into the dates, stirring constantly until date mixture thickens. Remove from heat and chill.

Prepare pastry as follows:
Pastry:
1/2 C oil
1 C honey
2 C whole wheat pastry flour
1 tsp salt
3 C rolled oats

Cream oil and honey. Add flour, salt and oats. Put half of the crumbled mixture in the bottom of an oiled and floured 9x13 inch baking dish. Spread with date mixture. Put the remaining half of the pastry on top of date mixture. Bake in moderate oven, 350°, for 40 minutes until golden brown. Cut into squares and serve hot, or let chill and serve with non-dairy whipped topping.

Pineapple Coconut Squares

1/2 C butter or vegetable oil
1/2 C honey
1 C whole wheat flour
1 tsp salt
1 C quick oats
1 C coconut

Cream oil and honey together. Add dry ingredients and crumble.

Filling:
3 C crushed pineapple
1/3 C corn starch

Put pineapple in small sauce pan and add cornstarch, stir constantly till it boils.

Put half of dough in oiled 9 x 13 pan and then put the filling on. Then sprinkle the rest of dough on top.

Bake at 325° for 20 to 25 minutes.

PIES

Whole Wheat Pie Crust

2 C whole wheat pastry flour*
1 tsp salt
5 T cold water
2/3 C lard, shortening or butter (lard makes the best pie crusts
   but is full of fat)

Sift flour and salt together. In separate bowl add to only 1/3
C of the flour, 5 T cold water. Cut in oil with remaining flour
to a medium-course texture (size of a pea). Add water and
flour mixture and stir with a spoon to form a ball. Roll out
and place in pie tins. Chill before baking. Bake 15 minutes or
till browned at 350°. Makes 3 single crusts.
* (Whole wheat pastry flour = substitute 2 T of cornstarch,
for 2 T of flour and sift together before using)

Whole Wheat Pie Crust with wheat germ

1 C whole wheat flour
1 tsp salt
1/2 C wheat germ
1/2 C oil
1/4 C cold soy milk (or water)
(Continued)

Mix dry ingredients. Pour oil and milk together in a cup but do not stir. Mix oil and milk with dry ingredients. Press into a ball then cut n half. Place each half of pastry between two 12-inch squares of waxed paper. Dampen table top to keep paper from slipping. Roll the crust between the paper, rolling from the center outward.

When rolled, remove the top paper, leaving the crust on the bottom piece of paper. Pick up the paper and flip crust right over the pie tin. Fit into pan, flute, trim. The crust is ready for baking. Bake 12 minutes or till browned at 400°.

Yield: 1 double crust or 2 single crusts.

Simple Pie Crust

2 C flour
1/4 tsp salt
1 C vegetable or olive oil
1/2 C water

Mix the flour and salt. Cut in oil and add water a little at a time. Roll out. Prick bottom and sides for single crusts. Bake 15 min at 375°. Makes 2 crusts.

Mock Graham Cracker Pie Crust

Made with Gluten Crunch
2 C Gluten Crunch (page 37) that has been put in blender for
   a finer texture
3 T butter
2 T flour
1/2 tsp cinnamon (optional)

Mix ingredients and press down in pie tin.
Bake in pre-heated 350° oven for 8 min. Cool and fill with
   favorite pie filling.

Makes 1 large pie crust

Note: You can replace up to 1 C of the Gluten Crunch with bran granules. Other fillings good with this crust are pecan cream, cheese cake, lemon custard, banana or coconut cream, and pineapple whip.

Whole Wheat Graham Cracker-Type Crust

1/4 C butter
1 T brown sugar
1 C whole wheat flour
1/2 tsp salt

Mix all ingredients. Press mixture into the bottom of a pie pan. Bake at 350° for 15 minutes. Makes 1 pie shell.

Sugarless Apple Pie

6-8 apples, sliced
one 12 oz can frozen apple juice concentrate
2 T cornstarch
Dash of cinnamon or cardamom (optional)
2 pie crusts, uncooked

Line pie tin with pastry, fill with apples.
Combine apple juice with cornstarch, then cook until thick.
Pour over sliced apples.
Top with crust. Bake 10 min at 375°, then 2 hours at 275°.

Honey Molasses Pumpkin Pie

Mix together:
3 C canned pumpkin
3/4 C honey
2 T molasses
1 T cinnamon
1 1/2 tsp ginger
1/4 tsp cloves, ground
4 eggs, slightly beaten
1 C evaporated milk

Mix well then pour into 3 unbaked shells. If desired, place pecan halves onto pastry shell before filling it. This helps prevent a soggy crust, especially if you are going to freeze it. Bake at 350° until knife comes out clean when inserted into center of pie, about 50 minutes.
Makes 3 pies

Pumpkin Pie

Mix together:
2 C scalded milk
1 C brown sugar

Add:
2 tsp pumpkin pie spices or 1/2 tsp each, cinnamon, ginger and nutmeg. Cook until thickened, then another 10 min, stirring frequently.

Add:
2 egg yolks, stir and cook 1 min.

Add:
1 1/2 C pumpkin
Cool and fill baked pie shell. Chill and top with whipped cream.

Yam Pie

1 1/3 C cooked, peeled yams (or pumpkin or squash)
1/2 tsp salt
1/3 C brown sugar
2 C crushed pineapple (do not drain)
3 T browned flour
1 T molasses
1/2 tsp vanilla
1 T butter
1/2 C soy milk powder
1/2 tsp grated orange rind
1/2 tsp ground coriander

Blend ingredients together thoroughly in blender. Pour into prepared, unbaked pie shell and bake 1 hour until done, 10 minutes at 400°, and finish at 350°.

Browned Flour

Toast flour till browned either by cooking in heavy non-oiled fry pan, stirring constantly or by placing on baking sheet at 250° for about 20 min.

DESSERTS

Apple Crisp

Mix together:
12 apples, pared, peeled and chopped
6 T whole wheat flour
1 tsp cinnamon
2 T lemon juice
1/2 tsp vanilla
1 C honey or sugar
Place mixed ingredients in a 9 x 12 oiled baking pan and top
   with mixture of:
1 1/4 C butter
3 C rolled oats
2 C brown sugar
1 C whole wheat flour

Bake for 1 hour at 350°.

Apple Pudding

1/2 C oil
1 1/2 C honey
2 eggs
4 C grated apple
2 C whole wheat flour
1 tsp salt
1 tsp soda
1 tsp cinnamon
1 tsp nutmeg
1/2 C chopped nuts

Cream oil, honey and eggs, beating until fluffy.
Add fruit. Stir in dry ingredients. Add nuts. Pour batter into
9" x 15" greased pan and bake at 350° for 45 min or until
tooth pick comes out clean. Serve warm or cold with sauce.

Sauce
1 1/4 C brown sugar

4 T butter
2/3 C dark Karo syrup
3/4 C evaporated milk

Boil to soft boil stage, cool, then beat in 3/4 C evaporated milk. Serves 10

Chocolate Orange Crunchies

1 C Gluten Crunch (page 37)
1/2 tsp powdered orange oil or 1/8 tsp orange extract
1/3 to 1/2 C mixed carob chips and mint chocolate chips

Place 1 C Gluten Crunch in saucepan to heat (over medium to high heat). Remove from heat. Stir in orange flavoring and chips. Place back on heat if chips are not completely melted.

Granola

7 C rolled oats
1 C each wheat germ, honey, oil, sesame seeds, sunflower
   seeds, unsweetened coconut, ribbon coconut, raisins, soy
   flour, chopped nuts.
1/4 C water
2 T vanilla
1 tsp almond extract
1 T salt

Mix together oats, wheat germ, honey, oil, sesame seeds, sunflower seeds, coconut, soy flour and chopped nuts. Mix together water, vanilla, almond extract and salt, then add this to the first mixture. Bake and stir in 300° oven until golden brown (about 30 minutes) stirring occasionally. Add raisins during last 10 minutes of baking.

Granola Bars: When making the whole recipe into bars double the liquids (oil and honey), mix well and press onto a cookie sheet to about 3/4 inch thick. Bake until brown, about 30 min (need not stir while baking.) Cut into bars and cool.

Graham Crackers

1/2 C evaporated milk
2 tsp lemon or vinegar
1 C oil
1/2 C honey
2 tsp vanilla
2 eggs
1 C raw or brown sugar
6 C whole wheat flour (to one of these Cups add 1 tsp each
   salt and soda)

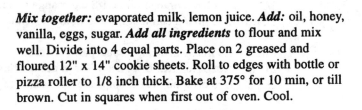

Mix together: evaporated milk, lemon juice. ***Add:*** oil, honey, vanilla, eggs, sugar. ***Add all ingredients*** to flour and mix well. Divide into 4 equal parts. Place on 2 greased and floured 12" x 14" cookie sheets. Roll to edges with bottle or pizza roller to 1/8 inch thick. Bake at 375° for 10 min, or till brown. Cut in squares when first out of oven. Cool.

Ideal for ice cream sandwiches; also for a graham cracker pie crust by crumbling the baked cracker and following any graham cracker pie crust recipe.

Gingerbread Crackers

2 C whole wheat flour
1 tsp baking powder
1/2 tsp soda
1 tsp ginger
1 tsp cloves
1 1/2 tsp cinnamon
1/ tsp nutmeg
1/4 tsp salt
1/2 C shortening
1/2 C honey or sugar
1/2 C molasses
1 egg yolk
Mix all ingredients together and roll out 1/4 inch thick. Use oil on rolling pin to prevent sticking. Cut with knife or cookie cutter for desired shapes. Bake on cookie sheet for 7 minutes at 350°.

CANDIES
Not all these treats are made from wheat but are wholesome and tasty.

Dried Fruit Treats

2 C softened dried fruit (this may include fruits such as dates, figs, prunes, cherries, apricots, raisins).
2 C unsweetened coconut
2 lbs dipping carob

Other ingredients that can be included (optional):
dry milk (non-instant)
protein powder
calcium powder
finely chopped seeds and nuts
carob powder

Put dried fruit through food chopper.
Mix with coconut or any other ingredient you desire (not dipping carob). If the combinations included do not stick together well, add honey until they do. On bottom of loaf pan, spread melted carob about 1/4 inch thick. Let cool until firm. Pat down on top of carob a layer of the fruit mix (about 1/2 inch thick).

Now spread another layer of the melted carob (1/4 inch thick) on top of the fruit mix. Let cool until firm enough to cut in rectangles or squares. Store in refrigerator.

Note: Carob has a taste similar to chocolate. It is an excellent food.

Hint: To soften fruit that has been dried too hard, put fruit in jar, run water in jar to top, then pour out immediately. Put lid on and store in refrigerator 12 to 24 hours. The fruit pieces will be chewy. If too moist, place out in the open until right consistency.

Coconut Balls

1/2 C honey
1 C powdered milk (non-instant)
1/4 tsp almond flavoring
1/4 C fine coconut

Roll into small balls and flatten on a Teflon or greased cookie sheet. Broil in oven until brown bubbles appear.

Peanut Butter Fudge

2 C peanut butter - plain or crunchy
2 C honey
2 C powder milk (non-instant)
Gluten Crunch (page 37)

Heat honey and peanut butter in heavy saucepan over low heat to a very soft, creamy texture, stir constantly or it will burn.
Pour into bowl of the powder milk and stir till thoroughly mixed. Use hands to knead in*. At this point you may add any dried chopped fruit, nuts, seeds, coconut, carob or chocolate chips, rice crispies, or Gluten Crunch. However, many children like it just plain.

Roll into balls or pat down in cookie sheets, baking pans, or anything flat with sides. Put in fridge or freezer till cool and slightly firm. Cut into squares with warmed knife. For continued firmness store in cool place.

* For a carmel-like texture beat with dough hook on electric bread mixer for about 5 minutes.

Note: For a more firm texture add 1/4-1/2 C more powder milk. The powder milk is not reconstituted. (No water added to it.)

Variations:
• Press an almond in each cut square ,chocolate chip etc. before putting in fridge.

• Roll small balls in toasted coconut
• Chopped dried fruit and seeds may be added
• Add carob powder to the basic mixture or baking cocoa
• 2 C rice crispies mixed in basic recipe is very good.

Peanut Butter Cups

Dip peanut butter balls into melted carob and press into crinkled candy cups. Let cool.

Old Fashioned Honey Candy

In a heavy sauce pan boil 2 C honey to the hard ball stage, 225° (when tested in cold water). Pour out on to an buttered surface. When cool enough to handle, butter hands and start pulling as you do for taffy. When honey holds its shape and is a golden color, twist into ropes and place back on to countertop or baking sheet and cut into pieces with scissors. Wrap individually with wax paper.

Variation:
2 C honey
1 C brown sugar
1 C cream
Combine honey, sugar and cream. Cook slowly to the hard ball stage and follow instructions as for "Old Fashioned Honey Candy"

Carob Crackle

1 lb dipping carob or chocolate
2 to 3 C Gluten Crunch (page 37)

Melt carob or chocolate in top of double boiler. The water in the pan underneath should be at medium heat. When carob is melted and you can stir it into a soft creamy texture, add Gluten Crunch. Stir in well. Spread out on cookie sheet or wax paper with back of spoon pressing to get the desired thickness. Score with knife when cool. Place in freezer at least 5 min for easy removal from pan. You could also drop

mixture with spoon onto sheet, which gives the haystack candy appearance.
Store in refrigerator.

Yield: 100 squares (1 1/2 in x 1 1/2 in x 1/4 in thick)

Cracker Jill

Syrup:
1 C water
1/2 C molasses (not blackstrap)
1/2 C honey

Dry Mixture:
1 1/2 C raw or roasted peanuts
1 C coconut, unsweetened or sweetened, wide or shredded)
1 C Gluten Crunch (page 37)

Cook ingredients for syrup to the soft ball stage.
Mix together the dry ingredients.
Stir both together till mixed well.
Pour out on oiled cookie sheets and bake 30 min at 250°. You will need to stir them around occasionally to get the melted liquid mixed with the nuts.
Take out of oven, stir again and let cool.
Break in pieces for serving.
Makes 11-12 C.

Replace or add to the dry mixture: sunflower, pumpkin or sesame seeds, oat or wheat flakes and other types of nuts. Raw almonds are very good in this.

Coconut Crisps

1 C shredded coconut
1/2 C whole wheat flour
1/2 tsp salt
1/4 C unbleached flour
water

Add enough water to other ingredients to make a firm dough.

Roll very thin, cut into wafers and bake in moderate oven — 350° until lightly brown.

Yield: approximately 2 dozen.

Marshmallows

Make your own—the healthful way. These honey marshmallows are simple to do and very tasty.

1/3 C water
1 1/2 T unflavored gelatin
3/4 C honey
1/4 tsp salt
1 tsp vanilla

Pour water in small saucepan and sprinkle gelatin on top. Let stand 5 min. Place pan over medium heat and stir only until dissolved. Do not boil. Add honey, salt and vanilla. Transfer all to a bowl and beat with electric beater until cool and tacky. Add nuts or dried fruit*. Sprinkle powdered milk or powdered sugar on bottom of 9" square cake pan. Pour mixture into pan and spread evenly. Let stand a few hours before cutting with hot knife. Roll squares in toasted coconut or dip in melted carob.

* If making Marshmallow Marvel candy, at this point add Gluten Crunch (page 37) in place of nuts and dried fruits. Spread on cookie sheet. Pat down even. Cool until firm enough to cut in squares.

Marshmallow Marvel

(similar to rice crispies treats)

40 marshmallows (10 oz pkg) or 1 C marshmallow creme or
   homemade marshmallow recipe
2 T butter
2 C Gluten Crunch (page 37)
1 C coconut (or 1 more cup Gluten Crunch)
(if using unsweetened coconut, add 2 T raw sugar or honey)

Melt marshmallows and butter in top of double boiler over
medium heat. When using marshmallow creme or honey
marshmallows there is no need for melting. Add Gluten
Crunch and coconut. Mix and prepare with any additional
variations.
Roll in balls and then in coconut, chopped nuts or Gluten
Crunch. A faster method is to pat down in buttered 9" square
pan, chill and when firm, cut in squares with a warmed knife.

Variations: Add 1 C any chopped nut meats, grated carob,
carob chips or chocolate chips, raisins or other dried fruits.
Or season with 1/2 tsp cinnamon
Hint. Moisten hands with water when forming balls. This
makes for faster and easier handling.

Gluten Surprise Balls

*The first bite is crunchy the rest is chewy
and full of flavor.*

Mix together:
1/2 C commercial Gluten Flour
1 tsp onion salt (omit for sweet variations)
1/4 tsp salt (or to taste)
Add and stir until flour has absorbed moisture
1/2 C water
Pull off or cut with scissors small 1/2 inch bits.
Place on greased cookie sheet.
Bake at 400° for 10-15 minutes.
YIELD: about 50 puffs
Store in paper bags.

Gluten Surprise Ball variations:

In 1/2 C water add seasoning of your choice, then add mixture to 1/2 C gluten flour and stir till all flour is moistened.

• Taco Flavor- 1 to 2 tsp taco seasoning mix

• Onion Flavor- 1 T onion mix recipe

• Sweet Snack- 2 tsp sugar, 1/2 tsp vanilla, 1/4 tsp cinnamon.

Other suggestions:

Roll in Parmesan cheese

Roll in Cinnamon Sugar

Gluten Licorice Pieces

A chewy, flavorful treat that lasts a long time. Prepared as in the gluten jerky simmering method but flavored with a variety of sweet flavors.

1 to 2 C raw, rolled out gluten cut into about 1/8 to 1/4 inch thick bite size pieces or strips (page 53)

Sweet Broth:

1 C water

1/2 C molasses

1/4 tsp black jel or paste food coloring (found at cake decorating stores)

2 to 3 T powdered Anise seed (according to your taste)

1/4 tsp salt

Drop raw gluten pieces into prepared broth and simmer for 5 minutes. If all the liquid has not simmered into the pieces, drain slightly and dehydrate. This can be done in your oven for about 30 minutes (page 53), in a dehydrator, or in the sun for 2 hours. Store in plastic bags. Pieces will become softer as they are stored.

Grape flavored pieces

1 C water

1/2 C honey or brown sugar

1/8 tsp violet or black jel or paste food coloring

2 T concentrated grape food flavoring ("Watkins" products)

1/4 tsp salt

Maple flavored pieces
1 C water
1/2 C honey or desired sweetener
2 T concentrated maple food flavoring
1/4 tsp salt

WHEAT SPECIALTIES

The wholesome grain, wheat, can be exciting, delicious and fun. This section opens up a whole new way of preparing wheat: breakfast meals, snacks, desserts, full meals, drinks, great breads, pastries and salads.

SPROUTED WHEAT

There are many ways to sprout. This simple method works well in sprouting wheat.

Place 3 T hard wheat in quart jar.
Cover with water. Let soak 6 hrs.
Place nylon or light screen on top of bottle and tighten down with ring.
Pour water off, rinse and pour off again.
Place bottle in dark, warm cupboard on its side.
The wheat should sprout in 2-3 days.
Rinse, drain and store in fridge.

A Sprout Treat

Chop 1/4 to 1/2 banana in dessert bowl.
Mix two parts wheat sprouts (not older than 2-3 days) to one part coconut.
Sprinkle about 1/3 C over banana. Then add a scoop of vanilla Ice cream on top.

Wheat Nut Balls

2 C sprouted wheat
1 C nuts
1 1/4 C bread crumbs
3/4 C milk
1 tsp salt
2 T oil
3/4 onion, chopped

Grind wheat sprouts, stir in remaining ingredients. Shape into small balls. Bake at 400° about 25-30 min. Top with sauce.

STEAMED WHEAT

In a steaming process, the wheat kernel becomes a very tender, easily digested food, as well as a very versatile addition to many recipes.

1 C whole kernel wheat (for faster cooking: soak overnight or at least 6 hrs in double amount of water; drain before steaming)
2 C water
1 tsp salt

Place wheat in water and salt in a 1 to 2 qt casserole dish or pan, uncovered, in a larger kettle for steaming. Place under the casserole an item that will keep the bottom of the dish from touching the kettle, for example, a canning jar ring. Add water to kettle and cover with just the one large lid.

Bring water in bottom of large pan to full rolling boil; boil for about 15 min. Reduce heat and simmer for about 2 hrs, if wheat has been soaked over night or 6 hrs. Simmer 3 to 4 hrs or until wheat is tender, if wheat has not been previously soaked. Refrigerate or freeze for storing. Makes 10 servings.

Suggestions for Serving
Heat and top with butter
Use in place of beans for Boston Baked Beans recipe
Serve topped with cream soup
Serve with vegetables such as celery, onion, bell pepper, etc.
Grind with cheese and butter. Bake.
Use in place of rice, in casseroles and custards.

Chili

4 C steamed wheat
3 C Ground Gluten pieces (page 17)
1 onion, chopped
1 clove garlic, minced
2 T cooking oil
2 tsp chili powder
1/2 tsp oregano
1 tsp salt
1 1/2 T whole wheat flour
5 C beef stock
1 C tomato sauce

Sauté Gluten or meat with onions and garlic until lightly browned. Add remaining ingredients and simmer over low heat 1 hr, stirring frequently. Serves 6.

Boston Baked Wheat

2 C steamed wheat
1 T Worcestershire sauce
1/4 C brown sugar or 1/4 C dark molasses
2 green onions, chopped
1 8 oz can tomato sauce
2 T soy bacon bits

Mix ingredients. Top with bacon bits. Bake covered 1 1/2 to 2 hours at 325 to 350°. Add Gluten Pieces just before serving. Serves 6.

CRACKED WHEAT
Cracked wheat is wheat kernels that have been coarsely ground.

Sift out any flour found in cracked wheat (common when home ground) by shaking it through a fine wire strainer. These fine particles can be used in other recipes or put with wheat kernels to be ground into flour.

1 C cracked wheat
1 3/4 C water
1 tsp salt (add after wheat has cooked)

Place water & cracked wheat in heavy sauce pan, bring to full boil, cover, turn off heat and let stand 15 minutes or till water is absorbed.

Cracked Wheat Cereal

Milk can be used in place of water when cooking over direct heat. Pour over cooked cereal either milk or fruit juice, and sweeten with honey, date sugar or fructose.

Cracked Wheat Salad

1 C sifted cracked wheat
2 C water
1/2 tsp salt
4 T mayonnaise or salad dressing
2 T chopped green pepper
1/4 C chopped green onions
3/4 C chopped celery
one 6 oz can tuna, flaked
Lettuce leaves
Parsley

Cook wheat. Cool. Place in large bowl. Mix in lightly the remaining ingredients. Place in lettuce-lined bowl. Garnish with parsley. Keep cool. Flavor is much improved if salad is made ahead of time and let chill several hours or overnight. Use this as a basic recipe. Many different combinations can be used with cracked wheat; ham, bacon bits, shrimp or corned beef. Try your favorite potato salad recipe, using cracked wheat in place of potatoes.

Cracked Wheat Balls

2 1/2 C cooked cracked wheat
2-3 lbs Cooked Ground Gluten Pieces (page 17)
2 eggs
1/2 C milk
3 T minced onion
1/4 tsp nutmeg
1 1/2 tsp salt
1/4 tsp Worcestershire sauce

Mix all together and form into balls. Roll in seasoned flour (page 39). Brown in large skillet with oil. Thicken 2 C seasoned broth. Just before serving add 1/2 C sour cream to wheat balls. Heat through. Serve on noodles, rice or potatoes. Yield: 70 balls

Cracked Wheat Patties

2 C cooked cracked wheat
1/2 C chopped onion
1 tsp soda (optional)
2 T dry non-instant milk
2 eggs, beaten
1/2 C fresh mint or parsley

Mix, season and fry in patties.

BULGUR WHEAT

Bulgur wheat is a pre-cooked and dried preparation of wheat, which gives the advantage of cooking faster than whole or cracked wheat.

It has a sweet, nut-like flavor and crunchy texture. If bulgur wheat is not available and you do not want to make your own, cracked wheat can be used in most recipes.

How To Prepare Bulgur Wheat

Method 1

Wash the whole-kernel wheat in cool water, then discard water.

Place wheat in medium saucepan and enough water to cover wheat about 2 in.

Bring to boil. Turn heat off, let rest 1 to 2 hrs.

Add more water if needed and bring to boil again, then let rest another 1 to 2 hrs.

Drain (use water for plants, soups or other cooking) and dry out in 200° oven until very dry. It can also be dried in dehydrator or in the sun on screen trays.

Method 2

Steam the washed wheat kernels in double the amount of water until the liquid is absorbed and the wheat is tender—about 1 hr or less. Spread thinly on cookie sheet or shallow pan and place in oven at 200° until it is dry enough to crack easily.

Remove chaff by rubbing the kernels between wetted hands. Crack the dried wheat in a mill or grinder to moderately fine, or use whole.

Store in airtight container on shelf.

To Reconstitute:

Boil 1 C bulgur to 2 C water for 5 to 10 min or soak overnight. It will double in volume.

Suggestions for Serving:

As a meat extender

Cooked and chilled, it can be added to cold salads, especially cole slaw Soaked overnight, it can be added to breads, rolls, and cookies Sauté onion, green pepper and a beef base. Add to reconstituted bulgur. Sauté this mixture until brown. Serve with cream of mushroom or chicken soup on top, then add chicken, pimentos, almonds, or tuna.

Bulgur Casserole

3/4 C bulgur, soaked in double amount water 2 hrs
2 T oil
1/4 C chopped onions
2 T chopped green bell pepper
1 C canned whole kernel corn
1 C sliced ripe olives
1 1/2 canned tomatoes
2 tsp lemon juice
1 tsp seasoned salt
2 tsp chili powder
1 garlic clove, minced
2 C Gluten Pieces (page 17)

Sauté onion and pepper. Combine with bulgur and remaining ingredients. Mix and place in 2 qt. casserole. Cook, covered in 350° oven for 1 hour. Add Gluten Pieces just before serving.

POPPED WHEAT

In heavy fry pan (no oil) over medium-high heat, pop (stirring and/or shaking constantly) 1 cup each of the following until lightly browned and popping sounds have almost stopped: Raw wheat kernels (soft wheat is most tender), pumpkin seeds (called Oriental or Pepita pumpkin seeds) and sunflower seeds. Mix well and add 1/2 tsp. oil with onion or garlic salt. Try adding soy sauce to moisten and sprinkle with chili powder and /or parmesan cheese. Herb Seasoning Salts are good. You may also want to add roasted nuts, raisins, carob chips, freeze dried peas, corn nuts or other trail mix items from your grocery store.

Parched Wheat
Use the Popped Wheat method for cooking.
Put 1/2 C or less wheat kernels in fry pan or heavy skillet (no oil). Turn heat on high, shaking or stirring continually. Let wheat get quite dark. Crack in blender or nut mill. Store in airtight container.

Broths and Gravies
Add parched wheat to liquids for darkening broths and gravies.

Hot Drink
Add 1-2 tsp to 1 C boiling water. Let steep, strain. Serve with milk and honey if desired.

For a variety in your wheat drink, add parched barley and rye to the parched wheat. This makes a blend similar to postum.

Homemade Noodles

1 C whole wheat flour
1 large well beaten egg
2 T milk
1/2 tsp salt

Combine flour, egg, milk and salt to make a stiff ball of dough. On a lightly floured surface roll dough until very thin, to about 18x20 inches. Let dry about 1-1/2 hours. Cut into 1/2 inch strips with a pizza or pastry cutter. (To save for future use; store in container that is not airtight.)
Drop into boiling soup or water. Cook 8 to 10 minutes.
Makes 8 ounces or 3 cups and the cost is about eleven cents.

BREADINGS

Breading and batter-dipping are used to protect foods to be fried, while giving them extra flavor.

Basic Batter

For skillet or deep frying

3/4 C whole wheat flour
1/2 C corn meal or white bean flour
1 T sugar
1/2 tsp baking powder
1/2 tsp salt
1/2 tsp chili powder or 1 T instant minced onions
2 eggs, beaten
1/2 C milk
1 T oil or melted butter

Mix dry ingredients together well. Add cornmeal. Combine to mixed remaining ingredients. Consistency should be like pancake batter. Additional milk may be needed.

Seasoned Coating Mix

Try this speedy, delicious seasoning for coating Gluten Steaks (page 15)

2 C fine dry bread crumbs
1/2 C flour
4 tsp salt
4 tsp paprika
2 tsp chicken seasoning (page 95)
1 tsp pepper
1/2 C oil

Combine bread crumbs, flour, salt, paprika, poultry seasoning and pepper in bowl. Mix thoroughly. Cut in oil until mixture resembles coarse crumbs. Place in covered container and store in a cool place. Makes 4 to 4 1/2 Cups.

Shake And Bake Mix

2 1/2 C fine dry bread crumbs (whole wheat best)
2 T dry onion soup mix (page 92)
1/2 tsp salt
1/2 tsp pepper
1/2 tsp garlic powder
1/2 tsp onion powder
1 T dried parsley flakes
1/2 tsp dried oregano flakes

Preheat oven to 325°. Coat each piece of gluten. Lay on cookie sheet. Drizzle a little butter over pieces and bake till browned.

Italian Seasoned Topping Or Breading Mix

1 C dry bran
3 T grated parmesan cheese
1/2 tsp Italian seasoning
1/4 tsp garlic powder

Brown the bran in an ungreased heavy skillet, stirring till browned and roasted. Cool and add remaining ingredients. Store in tightly covered container. Sprinkle on soups, salads, casseroles or use for a breading on gluten steaklets.

A section dedicated for your convenience. Information that will make cooking faster and easier with an emphasis on more wholesome substitutions.

MEASURES, WEIGHTS, AND EQUIVALENTS

MEASUREMENTS AND WEIGHTS

dash	= less than 1/8 tsp
1 teaspoon (tsp)	= 1/6 fl. oz.
1 1/2 teaspoons	= 1/2 T
3 teaspoons	= 1 T
2 tablespoons (T)	= 1/8 C
4 T	= 1/4 C
5 T + 1 tsp (2 2/3 ounces)	= 1/3 C
8 T (4 ounces)	= 1/2 C
12 T	= 3/4 C
14 T	= 7/8 C.
16 T	= 1 C
1 cup (C)	= 1/2 pt
2 cups (16 oz)	= 1 pint (pt)
2 pts (4 C)	= 1 quart (qt)
4 qts	= 1 gallon (gal)
16 oz	= 1 pound (lb)

Can Sizes

8 oz	= 1 C
#300	= 1 3/4 C
#1 tall	= 2 C
#303	= 2 C
#2	= 2 1/2 C
#21/2	= 3 1/2 C
46 oz	= 5 3/4 C
64 oz	= 8 C
#10	=12-13 C

SIMPLIFIED FOOD EQUIVALENTS

Apples	1 lb	3 medium (3 C sliced)
Bananas	1 lb	3 medium (2-1/2 C

sliced)

Dry Beans	1 lb(2 C) uncooked	6 C cooked
Butter	1 lb	2 C
	1/4 lb	1/2 C (l stick or 8 T)
	4 oz.	l stick
	1 oz.	2 T
Carob or chocolate	1 lb	1 C (melted)
Celery	2 stalks (chopped)	1 C

Cheese

American or cheddar	1 lb	4 C (grated)
	4 oz.	1 C (grated)
Cottage	1 lb	2 C
Cream	8 oz.	1 C (16 T)
	3 oz	6 T
Coconut	14 oz	5 1/3 C
	3 oz.	1 C

Cookie and Cracker Crumbs

Graham cracker	15-18 Squares	1 C
Graham cracker	3-1/2 oz(crushed)	1 C
Soda	7-9	1 C
Vanilla wafers	20-30	1 C
Zwieback	4-9	1 C

Cream

sour	16 oz	2 C
heavy	l pint	2 C (4 C whipped)

Dates (chopped)	7 oz.	1 C

Eggs

1 egg has the leavening power of 1/2 Tsp baking pwd.

whole	3 medium	1/2 C	
	2 large	3 small	
yolks	3	1/4 C	
	12-14	1 C	
whites		4 medium	1/2 C

Flour

All purpose	1 lb	4 C
Cake	1 lb	4-1/2 C (sifted)
Pastry	1 lb	4-1/3 C
Cornmeal	1 lb	3 C
Rye	1 lb	4- 1/2-5 C
whole wheat	1 lb	4-1/2 C

wheat, whole kernel	9 C	12 C
Lemon		
juice	1 lemon	2-3 T
peel	1 (grated)	1-1/2 tsp
Macaroni	1 lb (4-1/2 C)	8 C (cooked)
Marshmallows	1/4 lb	16 large
	1/4	2 C miniature
Molasses	8 fl.oz.	1 C
Nuts		
whole	1 lb. (shelled)	3-1/2 to 4 C
shelled	1/4 lb. (4 oz.)	1 C
chopped	5 oz.	1 C
Onion, medium	1 (chopped)	1/2 C
Orange		
juice	1 medium	1/4 to 1/3 C
peel	1 (grated)	1 T
Raisins	1 lb	3-1/4 C
Rice	1 lb (2 C raw)	8 C (cooked)
Sugar		
Brown	1 lb	2-1/4 C (packed)
Confectioners	1 lb	3-1/2 C
Granulated	1 lb	2 C.

Nuts and Fruits Most Commonly Used

	1 lb in Shell	1 lb Shelled
Almonds	3 1/2 C	1 to 1 3/4 C
Pecans	3 1/2 -4 C	1 1/4 C
Peanuts	3 C	2 1/4 C
Walnuts	4 C	1 2/3 C
Brazil	4 C	2 C
Chestnuts	4 C	1 1/2 C
Filberts	3 1/2 C	2 1/4 C

	Whole	Pitted
Dates	1 lb	2 1/4 C
Prunes	1 lb	3 C (uncooked)
Figs	1 lb	2 3/4 C
Raisins	15 oz	2 1/2 - 3 C
Candied Fruit	1/2 lb	1 1/2 C

SUBSTITUTIONS

If you don't have this	use this
1 tsp Baking Powder	1/4 tsp soda plus 1 tsp cream of tartar
1 C butter	1 C margarine or 7/8 C corn oil or 1/2 C powdered butter
1 C catsup or chili sauce	1 C tomato sauce & 1/4 C sugar & 2 T vinegar (for use in cooked mixtures)
1 sq chocolate (1 oz)	3 T carob or cocoa powder plus 1 T butter or oil
1 C corn syrup	1 C sugar & 1/4 C liquid
2 T cornstarch	2 T flour
1 egg	1/4 C water to which has been added 2 T soaked soy or garbanzo beans & liquefied or stir 2 T soy or bean flour to 1/4 C water (1/4 C=1large egg)
1 egg, whole	2 egg yolks plus 1 T water, esp. in baking or 2 egg yolks, for custards or 2 egg whites
1 T wheat flour, for thickening	1 T bean flour or 1/2 T cornstarch, arrowroot, gluten starch or 2 tsp quick cooking tapioca
2 C flour, cake or whole wheat pastry flour	1 3/4 C whole wheat flour or replace 2 T flour with 2 T cornstarch plus enough flour to make 2 cups
1 C all-purpose flour for baking breads	1/2 C bran or 1 C whole wheat flour
1 garlic, clove	1/8 tsp garlic powder
3 C Ground Gluten Pieces	1 lb ground meat

1 C honey	1 1/4 C sugar plus 1/4 C liquid
1 C whole milk	1/2 C evaporated milk plus 1/2 C water or 1 C water plus 1/4 C non-instant dry milk
cream (low-fat)	For cream-like consistency, mix one part powder to two parts water
1 C buttermilk (sour milk)	• 1 T lemon juice or vinegar plus enough evaporated, skim or low fat milk to make 1 C • 1 C plain yogurt • 1/2 C milk plus 1/2 C plain yogurt •1/4 C dry buttermilk powder mixed with 1 C water
1 C sour cream	Liquefy 1 C cottage cheese, 2 T Butter milk and 1/2- 1 tsp lemon juice or 1 C buttermilk for using in dressings or dips
1 C whipped cream	Beat until thick: 1/2 C non-fat pow-dered milk and 1/2 C ice water. Add 1/4 tsp vanilla, 1/2 tsp lemon juice and 1/4 C sugar or Vanilla yogurt (page 129)
1 oz. cream cheese	Liquefy 1 oz. ricotta cheese
1 C. mayonnaise	1 C plain yogurt or 3/4 yogurt mixed with 1/4 C mayon-naise
1 tsp mustard dry	1 T prepared mustard
Tabasco Substitute	(For 3-4 drops equivalent) use 1/8 tsp cayenne pepper
1 onion, small and fresh	1 T instant minced onion, rehydrated.
1 tsp onion salt	1 T dry onion flakes

1 C sugar

3/4 C honey minus 4 tsp
Use 1/2 C honey in baked products or
1 C firmly packed brown sugar.

1/4 C lemon juice

1 tsp powdered lemon juice + 1/4 tsp
water

1 C tomatoes, canned

1 1/3 chopped fresh tomatoes, simmered 10 min.

1 pkg yeast

1 T dry powdered yeast or 1 yeast
cake.

REPLACEMENTS

EGG - *Mix 2 C hot water and 4 T flax seed together and soak for 10 minutes. Liquefy in blender, for only 3 seconds (just enough to break the seeds open). Use 1 T of the blended mix in place of 1 egg. Use in Sauces, dressings and baking.

MILK - Soy milk, oat milk, rice milk, nut milks (see page 128-129), baby formulas, check with Health Food stores for more.

SWEETENERS - Honey, pure maple syrup, rice syrup, fruit syrup, powdered barley malt syrup or powder, sorgum, Sucanat, turbinado sugar, Stevia.

THICKENINGS - "ALL PURPOSE THICK JEL AND ULTRA GEL" A thickening agents which are also effective in reducing fats and replacing eggs in most recipes.
It is a modified food starch made from corn and is used in most of the commercial mixes found in your grocery store. Items such as dressing mixes, seasoning mixes, gravies, puddings, baking mixes and baby food. Whenever you see " modified food starch", or "starch", on the label it is this natural thickening product. It is like cornstarch but much better because it does not break down, go lumpy and separate when refrigerated or frozen and does not leave the after taste that corn starch does.
A MUST FOR EVERYONE who would like to reduce fat intake , cannot eat eggs or would like the convenience of faster and easier cooking.
Advantages:
* Replaces or reduces fat in dressings, white sauces, and baked goods.
* Freezer jam can be sugar free.
* Use only 2 T of the All Purpose Thick Jel to replace 4 egg yolks

Note: Those wanting more information and a cook book for this product plus another similar product called, "Ultra Jel", which thickens without heating, can write to:

Smith Distributing
9847 North 7200 West
Lehi, UT 84043

Please send a legal sized self addressed, stamped envelope

INDEX

Other Products Available from LM Publications

THE HERB WALK VIDEO

Learn how to identify 137 wild herbs for their edible and medicinal uses and outdoor safety. You will see the difference between look-alike poisonous plants. Gain confidence as you learn to recognize wild plants all around you. Clear,detailed identifying characteristics in full color. Manual which comes with the video is excellent for beginning botanists.

$29.95
With Free Manual

NATURE'S COMPLETE LIBRARY

The most comprehensive Natural Health Library on CD-ROM. 2,500 herbs, vitamins & minerals, 3,000 recipes, tinctures & formulas, 250 full-color illustrations, 11,500 cross-reference links and more. Search by author, book, words, phrases, word-stems, etc. Thesaurus and Synonym searches find words with similar meanings.

$69.95

Order Line 1 888 554-3727   **Voice/Fax** 801 374-1858
Credit Cards accepted S&H **$3.50** + **.50** for each additional item
Web Site http://www.vii.com/~herbsetc   **E-mail** herbsetc@mail.vii.com

THE WHOLESOME FOODS VIDEO

A convenient way to make low-fat, great tasting meals in just minutes. Five l5 min. Mini-classes. Whole wheat **Bread** making secrets. **Gluten** made easy. **Wheat** used in breakfast, main dishes, drinks and more. 3 minute **Cheeses** made from Powdered milk. Recipes and techniques for using **Beans** and bean flour to make soups and sauces, in minutes. Excellent for workshops or cooking classes in you own home. This video helps you prepare healthy, basic delicious foods.

$29.95 with
FREE Recipe booklet

THE AMAZING WHEAT BOOK

This unique cook book can save you time, money and health. Find delicious, wholesome, fast recipes that are easy enough for children to make. Perfect whole wheat breads, rolls and muffins in minutes, crackers, chips, soups, salads, ice cream and more. Learn how to make wheat meat in minutes at only a fourth the cost of meat (meatballs, jerky, veggie burger, sausage and even pet food.) Prepare your own seasoning mixes, saving money and eliminating chemical preservatives.

$15.95

Country Beans
by Rita Bingham

Soups, Sauces and Gravies in only 3 Minutes! 400 heart-healthy bean recipes. FAST guilt-free 3-minute Soups, Sauces and Gravies, and delicious fat-free 5-minute Bean Dips made using dry beans, peas or lentils ground to a flour in an ordinary wheat mill.

Forget what you've heard about the need to eliminate cream soups, sauces and gravies—now they're fat-free and good for you! Eat great meals to your heart's content—while lowering your cholesterol and reducing the risk of serious diseases! Most recipes are made without wheat or dairy products.Perfect for those with allergies and food intolerances.

$14.95

Natural Meals in Minutes
by Rita Bingham

Fed up with fat? Tired of the high cost of prepared foods? Learn to make delicious, low-fat, "save your life" meals with this new book in 3 sections:
- Introduction to Natural Foods
- Sprouting
- 3-Minute Powdered Milk Cheeses

Hundreds of recipes—low in fat, high in energy, completely nutritious, and so fast! Includes nutritional data for each recipe.

Wholesome beans, grains, nuts, seeds, fresh fruits and veggies, seasonings and spices combine to make great meals in 30 minutes or less, from home-style to gourmet! You'll find downright "sneaky" ways to incorporate natural foods into your everyday meals.

$14.95